Springer Theses

Recognizing Outstanding Ph.D. Research

Aims and Scope

The series "Springer Theses" brings together a selection of the very best Ph.D. theses from around the world and across the physical sciences. Nominated and endorsed by two recognized specialists, each published volume has been selected for its scientific excellence and the high impact of its contents for the pertinent field of research. For greater accessibility to non-specialists, the published versions include an extended introduction, as well as a foreword by the student's supervisor explaining the special relevance of the work for the field. As a whole, the series will provide a valuable resource both for newcomers to the research fields described, and for other scientists seeking detailed background information on special questions. Finally, it provides an accredited documentation of the valuable contributions made by today's younger generation of scientists.

Theses are accepted into the series by invited nomination only and must fulfill all of the following criteria

- They must be written in good English.
- The topic should fall within the confines of Chemistry, Physics, Earth Sciences, Engineering and related interdisciplinary fields such as Materials, Nanoscience, Chemical Engineering, Complex Systems and Biophysics.
- The work reported in the thesis must represent a significant scientific advance.
- If the thesis includes previously published material, permission to reproduce this must be gained from the respective copyright holder.
- They must have been examined and passed during the 12 months prior to nomination.
- Each thesis should include a foreword by the supervisor outlining the significance of its content.
- The theses should have a clearly defined structure including an introduction accessible to scientists not expert in that particular field.

More information about this series at http://www.springer.com/series/8790

Itia Amandine Favre-Bulle

Imaging, Manipulation and Optogenetics in Zebrafish

Doctoral Thesis accepted by
the University of Queensland, Brisbane, Australia

 Springer

Author
Dr. Itia Amandine Favre-Bulle
School of Mathematics and Physics
The University of Queensland
Brisbane, QLD, Australia

Supervisor
Prof. Halina Rubinsztein-Dunlop
The University of Queensland
Brisbane, QLD, Australia

ISSN 2190-5053   ISSN 2190-5061 (electronic)
Springer Theses
ISBN 978-3-319-96249-8   ISBN 978-3-319-96250-4 (eBook)
https://doi.org/10.1007/978-3-319-96250-4

Library of Congress Control Number: 2018948714

This Springer imprint is published by the registered company Springer Nature Switzerland AG
The registered company address is: Gewerbestrasse 11, 6330 Cham, Switzerland

Supervisor's Foreword

The field of laser micromanipulation which includes optical trapping in three dimensions, translation and rotation of nano- and micron-sized objects has found widespread application in a variety of fields such as quantum atom optics, biophotonics, biology and biomedicine. In laser micromanipulation, we use the fact that light can carry both linear and angular momentum which can be transferred to an object and cause its movement either only in a linear fashion or rotationally. In order to enable these sorts of transfer of momenta, we need to focus the beam using high numerical aperture objective to a diffraction-limited spot so that it will then have very high energy density and will be able to exert relatively strong force near its focus. Focusing of a laser beam results therefore in an irradiance so high that the fraction of transferred momentum to a small object is big enough to levitate it. This single-beam gradient trap, or laser tweezers, has been first demonstrated by Ashkin in 1986 [1]. Many recent applications of this phenomenon include holographic tweezers enabling trapping in 3D of many objects simultaneously, shaping light beams that are used for trapping to enable manipulation of larger objects as well as constructing innovative trap geometries, such as a reflection trap. The usefulness of these techniques has been demonstrated in many exciting quantitative studies in which the applied forces acting on the manipulated objects could be measured as well as displacements and applied torques. That led to unprecedented studies, for example, of uncoiling of DNA and the strength of the forces need for that, studies of molecular motors, development of microrheology and more. Typical forces that can be exerted by a single-beam gradient trap are of the order of pN, making manipulation of large objects questionable or needing further developments of the method. Apart from that, majority of laser micromanipulation is also normally restricted to shallow depth in the samples. Therefore, the use of laser micromanipulation in vivo poses some problems. For many highly relevant problems in biology, a combination of optical physics, behaviour, advanced microscopy and optogenetics can provide powerful way for answering questions that otherwise would be impossible to give an answer to. One of such fundamental problems is a study of the vestibular system, which controls balance and movement. The research community has dedicated a huge amount of effort towards understanding how external stimuli are converted into neural signals and how our brains process and interpret these raw signals to generate a neural representation of our surroundings. The vestibular system, which reports on gravity, head rotation and acceleration, remains comparatively poorly understood. Stimulating the vestibular system requires the rotation or acceleration of the subject, and this complicates or prevents many approaches for studying nervous system activity (electrophysiology, functional imaging or calcium imaging). This means that the vestibular field has a lot to gain if an approach can be developed for stimulating the vestibular system in a stationary preparation. In her Ph.D. studies, Itia Favre-Bulle made a seminal contribution towards studies of vestibular system in a stationary zebrafish. Trapping is a common technique, but to this point it has generally been done on objects up to a few microns across (DNA, proteins and synthetic beads for example) as a way of testing their optical properties or physical forces acting on them. By comparison, zebrafish otoliths are enormous (approximately 55 μm) and are optically heterogeneous. They are also 200 μm deep in a complex mix of tissues with different refractive indices. All of these factors complicated the goal of applying controlled, physiologically relevant forces to the otoliths. In her thesis, Itia performed meticulous mapping of the otoliths optical properties, theoretical modelling of the forces that were applied with the laser light, in vitro testing and finally attempts to trap the otoliths in vivo. This last step required a novel microscope capable of applying dual optical trapping (to enable manipulation of the two otoliths simultaneously) with synchronised imaging at two scales. In her thesis, Itia started off the work on this problem with an in-depth characterisation, both theoretical and experimental, of the interaction between light and brain tissue in order to quantify light scattered in the zebrafish brain tissue, which is of a great importance for optogenetics and any evaluations of the levels of light delivered deep into the brain tissue. These in turn help to quantify optical forces applied to the parts of the vestibular system. Using dual optical tweezers system in combination with imaging of the tail movements of the zebrafish as well as its eye movements, she was able to study in detail the response to fictive stimuli. Itia was then able to apply forces to otoliths in live immobile larvae and get logical behavioural responses from their eyes and tails, consistent with their compensation for perceived (fictive) acceleration or rotation. Since she was able to trap the left and right otoliths individually or together with a dual optical trap, and since she could adjust the power of the laser, this allowed her to map the ipsilateral/contralateral contributions that the otoliths make to the behaviours and the relationships that exist between the strength of the perceived stimuli and the graded responses that they elicit. In the final part of her dissertation, Itia combined fluorescence microscopy, imaging of the whole zebrafish and optical tweezers to visualise neuronal activity during the stimuli of the vestibular system and demonstrate which regions of the brain are involved in information processing. This thesis not only provides a detailed account of these very important studies but also gives an excellent review of the field and a necessary background to the methods that are developed for this specific study. It is a grand example of an interdisciplinary, highly innovative study that led to results that were impossible to attain before. This makes this thesis a highly recommended reading for both biologists and physicists who are interested in optics, biophotonics, optogenetics and vestibular system providing understanding how external stimuli are converted into neural signals and how our brains process and interpret these raw signals to generate a neural representation of our surroundings. It will pave the way for new and exciting developments in this interdisciplinary field.

Brisbane, Australia   Prof. Halina Rubinsztein-Dunlop
July 2018

Reference

1. A. Ashkin et al., Observation of a single beam gradient force optical trap for dielectric particles. Opt. Lett. **11**, 288 (1986)

Abstract

This thesis investigates the advantages and limits of using laser light for illumination and micromanipulation of the nervous system deep in vivo in zebrafish. Further interest in the study of neuronal pathways and functions in zebrafish led to employ laser micromanipulation for the stimulation of the systems of interest and the development of optogenetics system and light sheet microscopy for analyses of the processes. First, I present the investigation of the interaction of light and brain tissue as we aim to quantify light scattered in zebrafish brain tissue with depth. This is of great importance for optogenetics as it uses light to control (turn on or off) neurons and the unwanted/scattered light out of the aimed region could significantly affect results of experiments. The measurements and model developed to predict light propagation through brain tissue and how much a focussed beam broadens and alters with depth are presented in detail. Results show that illumination of individual neurons is possible at depth in the zebrafish brain, despite scattering that results from shallower neural tissue. This means that the approach presented allows for optogenetics manipulation of single neurons to be performed without optical correction for scattering. Secondly, I construct an optogenetics system where we employ advanced optics and novel algorithm to deliver versatile two-dimensional illumination that can be used at any depth in the fish. I present an example of a problem that I have solved and resulting optogenetics results that were obtained, using this new technology. Lastly, I investigate another facet of light's interactions with biological systems by performing optical trapping of the ear stones (otoliths) in live zebrafish larvae. The anterior ear stone, called the utricular otolith, has been found to be responsible for sensing acceleration. It is a massive calcium carbonate crystal located deep in the brain. To be able to manipulate it without affecting the rest of the animal, I used optical trapping with a focussed infrared laser beam. Using optical traps, I have been able to apply forces in different directions in a highly controlled way to simulate acceleration and to look at the reaction of the fish depending on the amplitude and direction of the force. Results show that the tail and the eyes are directly responsive to the acceleration sensory system. Finally, I combined different optical systems to visualise neuronal activity in the whole zebrafish brain with optical trapping and show which regions are involved in the information processing. Determining the brain regions involved in acceleration sensing is of great importance, and further investigations need to be done to determine what circuits are necessary for the detection of and behavioural response to acceleration stimuli.

Publications Related to this Thesis

I. A. Favre-Bulle, D. Preece, T. A. Nieminen, L. A. Heap, E. K. Scott, and H. Rubinsztein-Dunlop. *Scattering of sculpted light in intact brain tissue, with implications for optogenetics*. Scientific Reports **5**, 11501 (2015).

I. Favre-Bulle, T. A. Nieminen, D. Preece, L. A. Heap, E. K. Scott, and H. Rubinsztein-Dunlop. *Computational modeling of scattering of a focused beam in zebrafish brain tissue*. In *Optics in the Life Sciences*, OSA Technical Digest (online), p. JT3A.34 (Optical Society of America, 2015).

I. A. Favre-Bulle, A. B. Stilgoe, H. Rubinsztein-Dunlop, and E. K. Scott. *Optical trapping of otoliths drives vestibular behaviours in larval zebrafish*. Nature Communications **8**, 630 (2017).

Acknowledgements

I could not have come this far and finish this Ph.D. without the help and support from many people that I would like to acknowledge.

Above all, I would like to thank my parents, Jean-Claude and Dominique, for their unconditional love and support which could move mountains. My partner, Nicolas for his massive support and great patience at all times.

I am grateful for the amazing opportunity that my supervisors Halina and Ethan have given me but also for their great guidance and understanding through out the entire time. I would also like to thank Alex, Gilles and Timo for their great help in the work I presented, Daryl and David for their support, help, and for sharing their experiences at diverse stages of my Ph.D. Every member of the two groups I have been working in made my Ph.D. a fun place to live in so thank you Lena, Anatolii, David, Emmanuel, Ann, Michael, Jamie, Neha, Sarah, Lachlan and Isaac.

I have special thanks to Shu, Swaantje and Lucy for our philosophical discussions, the sharing of our experiences in the good and bad moments of our PhDs and their support.

Finally I have some friends that I would like to thank for their great support without necessarily realising it: Mari-wenn, Estelle and Arnaud.

Contents

Abbreviations

The following list is neither exhaustive nor exclusive, but may be helpful.

BFP	Back focal plane
ChR2	Channelrhodopsin
DAPI	4',6-diamidino-2-phenylindole
DM	Dichroic mirror
dpf	Days post-fertilization
ETL	Electrically tunable lens
FP	Fluorescent protein
FPS	Frame per second
FWHM	Full width at half maximum
GECI	Genetically encoded calcium indicator
GM	Gimbal mirror
GS	Gerchberg–Saxton
HWP	Half-wave plate
IR	Infrared
LMA	Low melting agarose
LUT	Lookup table
MO	Microscope objective
NA	Numerical aperture
NpHR	Halorhodopsin
OL	Offset lens
OT	Optical trapping
PBS	Polarising beam splitter
PSD	Position-sensitive detector
PVL	Periventricular layer

RL	Relay lens
SLM	Spatial light modulator
SPIM	Selective planar illumination microscopy
TL	Tube lens
VOR	Vestibulo–ocular reflex

Chapter 1
Introduction

When aiming to understand the brain, its complex network and constitution, one has to develop tools to visualise and control the information flowing through it on a cellular level. Methods for studying the brain are flourishing through the development of many new experimental methods that are able to give answers to long-standing problems.

For years scientists have developed systems and methods to study the brain on numerous models: mouse [1, 2], Drosophila [3, 4], and C. elegans [5], to name a few. In this thesis, I will discuss the development and application of such tools for the use in the zebrafish model system.

Zebrafish is widely used as a complex model organism as it has a brain structure similar to other vertebrates including mammals but the brain is comparatively small and therefore easier to study than in mammals (Fig. 1.1). Both larval and adult zebrafish are currently studied to improve our understanding of brain function and dysfunction [6, 7]. They are also used to model various brain disorders and study their behaviour and drug responses for neuropharmacology [7, 8]. The main advantages of larvae zebrafish as a model organism are that they develop externally [9] and are optically transparent. Despite their small brain size, larvae at 5 days postfertilisation (dpf) have functioning sensory systems for visual [10–12], auditory [13–15], somatory [16, 17], water flow [18–20], and vestibular stimuli [21, 22]. These make them a ideal system for sensory studies on the cellular level.

On the cellular level however, few methods allow to visualize neuronal activity in real time. Two examples are electro-physiology and fluorescent imaging. Since neurons communicate and deliver their signals through electric potentials, numerous studies use electro-physiology to visualize neuronal activity. It uses an electrode, placed in the region of interest, and measures the flow of ions which gives the measure of the potential and therefore the activity of that region. This is a very common technique for studying electrical signals of nervous but also other bodily activity. The particularity of this technique is that one can test and measure electrical activity from a single cell to an entire organ. The activity of a neuron is also accompanied with

© Springer International Publishing AG, part of Springer Nature 2018
I. A. Favre-Bulle, *Imaging, Manipulation and Optogenetics in Zebrafish*,
Springer Theses, https://doi.org/10.1007/978-3-319-96250-4_1

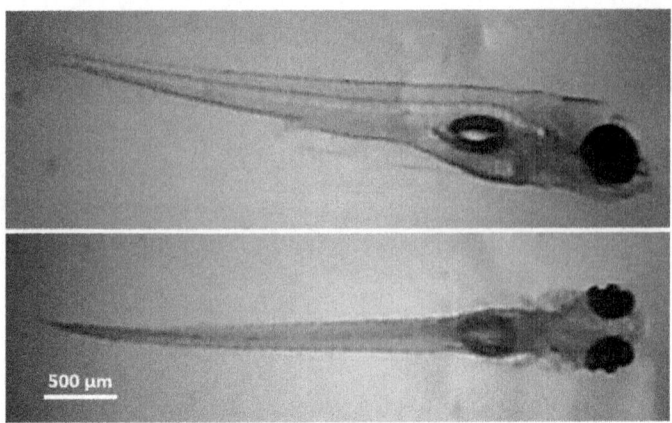

Fig. 1.1 Top and side view of a 6 days post-fertilization (dpf) larvae zebrafish

a peak of calcium ions concentration, within its body. Which implies that neuronal activity can also be visualized through the variation of calcium ions concentration. Recently, calcium imaging has emerged and various calcium indicators, based on fluorescence, were developed and designed to be expressed in various regions of the brain. By expressing the fluorescent calcium indicators in the whole brain, and imaging the fluorescent light, one can visualize and study the complex network in real time in response to various stimuli.

In addition, researchers recently found light sensitive ion channels [23, 24] that genetically modified cells can express on their membrane surface. Those channels can therefore allow the transfer of ions inside the cell resulting in the control of its activity or inhibition of activity. Those proteins, called opsins [25–28], have the particularity to be controlled with light, which implies that optical systems able to deliver light precisely to the cells of interest, need to be developed. This new method of visualizing cells networking is called optogenetics [29–32]. Optogenetics, asks to first, express a protein in the cells of interest to sensitize them to light, and second, develop an optical system to deliver light very precisely to those specific cells in order to control them. Tapping into the full utility of optogenetics, therefore, requires the development of specialised optical tools for visualising fluorescence signals or for sculpting light into desired shapes necessary for exciting prescribed sets of neurons.

Since brain tissue adsorbs and scatters light, it is essential to account for light distortion in the optical systems designs in order to precisely target the cells of interest. Zebrafish embryos can be genetically modified to be transparent, which facilitates the transmission of light trough out the whole brain. However, its brain is constituted of dense and sparse regions which can variably distort the light.

This leads us to the first part of this thesis (Chaps. 2–4) where I aim to understand how light propagates through brain tissue to establish and understand the possible losses of light through scattering. I also refine pre-existing optical systems used in these studies to ensure the optimal geometry of the delivered light. I then demonstrate how such sculpted light can be used to manipulate neural activity in targeted regions of the brain.

Analysis and understanding of light scattering is a very difficult question for inhomogeneous and permanently changing media such as living animals. Scattering occurs when light encounters a surface of a different refractive index, and these surfaces are numerous in an animal. This is also the case for a single cell of a few microns in size as the cell itself is heterogeneous. These scattering events result in a permanent change of light transmission over time within an animal or cell (Fig. 1.2).

The impacts of biological tissues on light are therefore notoriously difficult to model. Even simplified models [33, 34], in which neurons comprise only cell nuclei and membrane are very time consuming for calculating wavefront distortions through 10–20 microns of tissue. Given that a larval zebrafish brain is 200–300 microns thick, a calculation of wavefront distortion is not feasible. For this reason, I have generated a simplified model for the general scattering of light through tissue, based on the Monte Carlo method.

Monte Carlo method is a computational algorithm which uses random parameters or inputs to solve complex problems. In the case of light scattering, the random parameters would be scattering events locations and directions, which are dictated by the scattering phase function defined by the model.

A simplified model, for the general light scattering in tissue, is to consider only photons going through a homogeneous media, with a scattering coefficient calculated based on the density of neurons per volume and the wavelength of light used for the experiments. In Chap. 2, I present the Monte Carlo method, light scattering model and the details of the equations included in my code, used to solve the complex problem of light scattering. This code uses the defined scattering coefficients to determine the probability for a photon to be transmitted, scattered or reflected at every step of

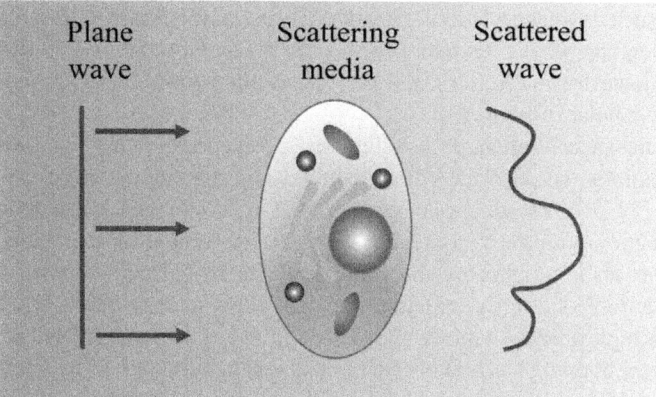

Fig. 1.2 Sketch of a light wavefront distorted by a cell. The refractive index heterogeneity of a cell distort light wavefronts significantly and affect illumination or imaging

the calculation. I show that it takes only few hours to simulate the path of 4 millions photons.

In Chap. 3, I present the set-up we built to support and verify this theory with in-vivo measurements. I present the results of backscattered light in zebrafish brain at different depth and in different brain regions and show very similar results to the Monte Carlo code developed in Chap. 2.

Finally, based on the approximation of minimal distortions at low depth provided in this chapter and Chap. 2, in Chap. 4 we designed and built an optogenetics system able to create flexible patterns of illumination. To shape the light in order to target precisely very specific regions of the brain, we use a spatial light modulator (SLM). Our SLM is a high definition display where each pixel can modulate the incoming light by reflection on their surface. It can therefore modulate the light in one plane (the screen plane) by displaying patterns and modify light propagation and location. This new system allows to excite targeted neurons, thereby studying their specific contributions to broader circuits. In Chap. 4, I also give an example of experiments we performed, using such a system, and the answers we obtained.

From Chaps. 2–4, we gained a better understanding of the light scattering means which led us to use light based methods for further quantitative investigations of the neuronal networks, leading to a better and deeper understanding of the overall system.

The second part of this thesis (Chaps. 5–6) focusses on gaining further understanding of light interaction in biological systems and looking at the very specific system of the vestibular system in zebrafish and more specifically the utricular otolith (or ear–stone) responsible for acceleration detection. The utricular otolith, located in the each ear of the zebrafish, is a calcium carbonate crystal built by the star-maker protein [35]. The utricular otolith has an ellipsoidal shape and is attached to the ear membrane with hair cells which are themselves connected to the ear-stone ganglia. Past studies focussed on the formation of such structures in the ear trying to answer questions such as what proteins are involved in their formation, how fast the ear-stone grows, when and what is essential to its growth etc. However, very little is known about the downstream circuitry, and how the otolith's movements are interpreted to produce vestibular responses.

To simulate acceleration, past studies have developed systems that accelerate, not only the otoliths, but also the whole fish body by rotating or tilting it in different directions [22, 36]. These types of activation of the vestibular system do not achieve pure interaction with the vestibular system. The movement of the subject in these preparations also complicates studying vestibular neural circuits with techniques such as electrophysiology or calcium imaging. However, as otoliths are transparent and have a high refractive index, they could in theory be manipulated with optical traps. Optical trapping has first been demonstrated by Ashkin [37] by levitating latex spheres in water with a focussed beam of light. Transparent particles with a higher refractive index than its surrounding receive and scatter light in a different direction causing a non zero total momentum on the particle which creates a force and therefore a movement. It has been shown that particles are attracted towards the highest intensity regions (by what is commonly called gradient force) and also experience a force in the beam propagation direction (commonly called scattering force). Since its demonstration, optical trapping has been improved, further developed and integrated into biological systems [38] for the measurement of visco-elastic properties [39], manipulation of cells or proteins [40, 41], drug delivery [42], studies of molecular motors [43, 44].

In Chap. 5, we investigate the optical properties of otoliths, questionning the possibility of their manipulation with optical traps experimentally and theoretically. Freeing them in water, I showed that we can manipulate and move free utricular otolith in X and Y but not Z, mostly due to their weight. We building a theoretical model of the distribution of forces within the structure to identify the regions where maximum forces can be applied. Our use of optical trapping here is to help to solve the problem of acceleration studies in-vivo. I first performed manipulation of free otolith in water and we evaluated the theoretical and experimental forces which can be possibly applied. I also determined the optimal regions in otoliths for optical traps to exert maximum force when manipulating in-vivo. Placing an optical trap on the edge of an otolith will be equivalent to the application of forces in chosen directions and, for the fish, this should simulate acceleration. In addition to optical manipulation, we also want to measure behavioural and neuronal responses by combining our optical trapping system with existing optical imaging systems. We therefore investigated existing combination of systems, their advantages and drawbacks for our application leading to the final construction of our own combined system.

In Chap. 6, I present the behavioural response to in-vivo manipulation of the anterior otoliths and present how eye and tail motions have very specific patterns of movements. The results from these experiments show clearly and consistently that the targeting of the lateral edge of one utricular otolith results in the deflection of the tail in the contralateral direction. Furthermore, the overall tail deflection response to a bilateral stimulus appeared to be the linear combination of the two otoliths' independent contributions. Similarly, the imaging of the eyes motion shows compensatory movements as a simultaneous rolling of both eyes. The consistent behavioural responses to otolith manipulation firmly establish that movements of the utricular otoliths alone are sufficient to drive compensatory responses across the body of a larval zebrafish.

In the last section of Chap. 6, I present my preliminary results of the brain activity in response to the utricular otolith manipulation. I show how I implemented two selective planar illumination microscopes (SPIM) in the optical system to do fluorescence imaging and look at neuronal activity in the whole brain with cellular resolution. Coupled with an electrical tunable lens (ETL), these planes allow to illuminate a single plane within the fish at a variable depth without fish movements. Results show responses to utricular otolith manipulation in regions believed to be involved in motion such as the medial longitudinal fasciculus (nMLF) [45] or the medial octavolateralis nucleus (MON) [46], and in several other regions. These promising results are currently being further analysed to have a precise and clear mapping of the vestibular information processing.

Finally, in Chap. 7, I conclude on the implications of the results obtained and on the future possible studies.

References

1. T.N. Seyfried, G.H. Glaser, A review of mouse mutants as genetic models of epilepsy. Epilepsia **26**(2), 143 (1985)
2. A.P. Arnold, X. Chen, What does the four core genotypes mouse model tell us about sex differences in the brain and other tissues? Front. Neuroendocrinol. **30**(1), 1 (2009)
3. K. Rein, M. Zockler, M.T. Mader, C. Grubel, M. Heisenberg, The drosophila standard brain. Curr. Biol. **12**(3), 227 (2002)
4. M.B. Feany, W.W. Bender, A drosophila model of parkinson's disease. Nature **404**(6776), 394 (2000)
5. C.H. Rankin, C.D.O. Beck, C.M. Chiba, Caenorhabditis elegans: a new model system for the study of learning and memory. Behav. Brain Res. **37**(1), 89 (1990)
6. W. Norton, L. Bally-Cuif, Adult zebrafish as a model organism for behavioural genetics. BMC Neurosci. **11**(1), 90 (2010)
7. A.V. Kalueff, A.M. Stewart, R. Gerlai, Zebrafish as an emerging model for studying complex brain disorders. Trends Pharmacol. Sci. **35**(2), 63 (2014)
8. P. McGrath, C.-Q. Li, Zebrafish: a predictive model for assessing drug-induced toxicity. Drug Discov. Today **13**(9–10), 394 (2008)
9. C.B. Kimmel, W.W. Ballard, S.R. Kimmel, B. Ullmann, T.F. Schilling, Stages of embryonic development of the zebrafish. Dev. Dyn. **203**(3), 253 (1995)
10. J.S.S. Easter, G.N. Nicola, The development of vision in the zebrafish (danio rerio). Dev. Biol. **180**(2), 646 (1996)
11. J. Chhetri, G. Jacobson, N. Gueven, Zebrafish-on the move towards ophthalmological research. Eye **28**(4), 367 (2014)
12. G. Gestri, B.A. Link, S.C. Neuhauss, The visual system of zebrafish and its use to model human ocular diseases. Dev. Neurobiol. **72**(3), 302 (2012)
13. A.A. Bhandiwad, D.G. Zeddies, D.W. Raible, E.W. Rubel, J.A. Sisneros, Auditory sensitivity of larval zebrafish (danio rerio) measured using a behavioral prepulse inhibition assay. J. Exp. Biol. **216**(Pt 18), 3504 (2013)
14. Q. Yao, A.A. DeSmidt, M. Tekin, X. Liu, Z. Lu, Hearing assessment in zebrafish during the first week postfertilization. Zebrafish **13**(2), 79 (2016)
15. T.T. Whitfield, Zebrafish as a model for hearing and deafness. J. Neurobiol. **53**(2), 157 (2002)
16. A.M. Palanca, S.L. Lee, L.E. Yee, C. Joe-Wong, L.A. Trinh, E. Hiroyasu, M. Husain, S.E. Fraser, M. Pellegrini, A. Sagasti, New transgenic reporters identify somatosensory neuron subtypes in larval zebrafish. Dev. Neurobiol. **73**(2), 152 (2013)
17. M. Haesemeyer, D.N. Robson, J.M. Li, A.F. Schier, F. Engert, The structure and timescales of heat perception in larval zebrafish. Cell Syst. **1**(5), 338 (2015)
18. A.W. Thompson, G.C. Vanwalleghem, L.A. Heap, E.K. Scott, Functional profiles of visual-, auditory-, and water flow-responsive neurons in the zebrafish tectum. Curr. Biol. **26**(6), 743 (2016)
19. W.J. Stewart, G.S. Cardenas, M.J. McHenry, Zebrafish larvae evade predators by sensing water flow. J. Exp. Biol. **216**(3), 388 (2013)
20. M.J. McHenry, K.E. Feitl, J.A. Strother, W.J. Van Trump, Larval zebrafish rapidly sense the water flow of a predator's strike. Biol. Lett. **5**(4), 477 (2009)
21. B.B. Riley, S.J. Moorman, Development of utricular otoliths, but not saccular otoliths, is necessary for vestibular function and survival in zebrafish. J. Neurobiol. **43**(4), 329 (2000)
22. S.J. Moorman, R. Cordova, S.A. Davies, A critical period for functional vestibular development in zebrafish. Dev. Dyn. **223**(2), 285 (2002)
23. M.R. Banghart, M. Volgraf, D. Trauner, Engineering light-gated ion channels. Biochemistry **45**(51), 15129 (2006)
24. J.J. Chambers, R.H. Kramer, Light-activated ion channels for remote control of neural activity. Methods Cell Biol. **90**, 217 (2009)

25. G. Nagel, T. Szellas, W. Huhn, S. Kateriya, N. Adeishvili, P. Berthold, D. Ollig, P. Hegemann, E. Bamberg, Channelrhodopsin-2, a directly light-gated cation-selective membrane channel. Proc. Nat. Acad. Sci. **100**(24), 13940 (2003)

26. F. Zhang, L.-P. Wang, E.S. Boyden, K. Deisseroth, Channelrhodopsin-2 and optical control of excitable cells. Nat. Methods **3**(10), 785 (2006)

27. A.B. Arrenberg, F. Del Bene, H. Baier, Optical control of zebrafish behavior with halorhodopsin. Proc. Nat. Acad. Sci. **106**(42), 17968 (2009)

28. B. Schobert, J.K. Lanyi, Halorhodopsin is a light-driven chloride pump. J. Biol. Chem. **257**(17), 10306 (1982)

29. O. Yizhar, L.E. Fenno, T.J. Davidson, M. Mogri, K. Deisseroth, Optogenetics in neural systems. Neuron **71**(1), 9

30. K. Deisseroth, Optogenetics. Nat. Methods **8**(1), 26 (2011)

31. L. Fenno, O. Yizhar, K. Deisseroth, The development and application of optogenetics. Annu. Rev. Neurosci. **34**, 389 (2011)

32. R.T. LaLumiere, A new technique for controlling the brain: optogenetics and its potential for use in research and the clinic. Brain Stimul. **4**(1), 1 (2011)

33. A.N. Timo, L.Y.L. Vincent, B.S. Alexander, K. Gregor, M.B. Agata, R.H. Norman, R.-D. Halina, Optical tweezers computational toolbox. J. Opt. A Pure Appl. Opt. **9**(8), S196 (2007)

34. C. Linton, Electromagnetic scattering by particles and particle groups: An introduction. Contemp. Phys. **56**(3), 402 (2015)

35. C. Sollner, M. Burghammer, E. Busch-Nentwich, J. Berger, H. Schwarz, C. Riekel, T. Nicolson, Control of crystal size and lattice formation by starmaker in otolith biomineralization. Science **302**(5643), 282 (2003)

36. W. Mo, F. Chen, A. Nechiporuk, T. Nicolson, Quantification of vestibular-induced eye movements in zebrafish larvae. BMC Neurosci. **11**(1), 1 (2010)

37. A. Ashkin, Acceleration and trapping of particles by radiation pressure. Phys. Rev. Lett. **24**, 156 (1970)

38. A. Ashkin, J.M. Dziedzic, Optical trapping and manipulation of viruses and bacteria. Science **235**(4795), 1517 (1987)

39. J.S. Bennett, L.J. Gibson, R.M. Kelly, E. Brousse, B. Baudisch, D. Preece, T.A. Nieminen, T. Nicholson, N.R. Heckenberg, H. Rubinsztein-Dunlop, Spatially-resolved rotational microrheology with an optically-trapped sphere. Sci. Rep. **3**, 1759 (2013)

40. A. Ashkin, J.M. Dziedzic, T. Yamane, Optical trapping and manipulation of single cells using infrared laser beams. Nature **330**(6150), 769 (1987)

41. Y. Pang, R. Gordon, Optical trapping of a single protein. Nano Lett. **12**(1), 402 (2012)

42. C. Rusu, R. van't Oever, M.J. de Boer, H.V. Jansen, J.W. Berenschot, M.L. Bennink, J.S. Kanger, B.G. de Grooth, M. Elwenspoek, J. Greve, J. Brugger, A. van den Berg, Direct integration of micromachined pipettes in a flow channel for single dna molecule study by optical tweezers. J. Microelectromech. Syst. **10**(2), 238 (2001)

43. S.M. Block, B.J. Schnapp, L.S.B. Goldstein, Bead movement by single kinesin molecules studied with optical tweezers. Nature **348**(6299), 348 (1990)

44. H. Kojima, E. Muto, H. Higuchi, T. Yanagida, Mechanics of single kinesin molecules measured by optical trapping nanometry. Biophys. J. **73**(4), 2012 (1997)

45. T.R. Thiele, J.C. Donovan, H. Baier, Descending control of swim posture by a midbrain nucleus in zebrafish. Neuron **83**(3), 679 (2014)

46. J.C. Liao, M. Haehnel, Physiology of afferent neurons in larval zebrafish provides a functional framework for lateral line somatotopy. J. Neurophys. **107**(10), 2615 (2012)

Chapter 2
Light Scattering in Brain Tissue Using Monte Carlo Method

As discussed in the Introduction, optogenetics uses light to drive or visualise neural activity. Being able to quantify and predict how light is scattered is crucial for optogenetics since light delivered to the neuron will determine the activity of neurons. Any superfluous light, delivered to the neurons that were not meant to be there for the particular study, will lead to erroneous results. As brain tissue is a non-negligible scattering medium, beams are altered and spread with depth through neural tissue. To quantify this process, in order to make sure that we know which neurons are targeted or potentially affected, I implemented a Monte Carlo method which demonstrates and quantify loss of intensity of focused laser beams with depth. This method and model have been summarised and published in:

I. A. Favre-Bulle, D. Preece, T. A. Nieminen, L. A. Heap, E. K. Scott, and H. Rubinsztein-Dunlop. *Scattering of sculpted light in intact brain tissue, with implications for optogenetics*. Scientific Reports **5**, 11501 (2015), [1].

2.1 Monte Carlo Method and Our Model

2.1.1 Introduction to Monte Carlo

Monte Carlo method is a computational algorithm which is based on the idea of using random parameters or inputs to solve complex problems. The method was developed in 1940 by scientists who named it after the city in Monaco famous for its casinos and games of chances. Their initial problem was to determine the neutron diffusion accompanying the explosion of the atomic bomb. As solving the problem analytically was impossible with their existing technology, they had to solve it numerically and Monte Carlo method was found to be surprisingly effective in calculating the solution to the problem.

© Springer International Publishing AG, part of Springer Nature 2018
I. A. Favre-Bulle, *Imaging, Manipulation and Optogenetics in Zebrafish*,
Springer Theses, https://doi.org/10.1007/978-3-319-96250-4_2

Nowadays Monte Carlo method is applied to a wide range of problems and applications in science, engineering, finance and business. It proves to be very useful when solutions to a particular problem are too complex to compute analytically.

2.1.2 Complexity of Models and the Monte Carlo Solution

To be able to precisely determine how light is affected when passing through a defined medium, one has to determine how each element constituting this medium affects the light transmission. In the case of brain tissue, we would have to consider each neuron, its organelles, their different shapes and sizes, and their refractive indices [2, 3] to have the exact result of light transmission. However, this problem is far too complex to calculate and it is essential to make approximations on the geometry and number of elements considered to get results with reasonable calculation times.

As nuclei are relatively large (few microns in diameter), they scatter predominantly at small angles from the beam propagation direction, which results in a gradual spreading of a focussed beam and scattered photons to stay in a small region around the focal spot. Small scatterers present in the cell, such as organelles, tend to scatter isotropically[1] and are much less likely to affect the region around the focal spot. For these reasons, I have ignored the effects of smaller scatterers in my model and considered only scattering by nuclei. This led me to a very simple model in which I simulated scattering in a medium composed of randomly distributed particles (assumed as spheres) embedded in a medium of lower refractive index [4, 5].

As a first approximation, we considered nucleus to be uniformly spherical. Despite this rough approximation, our results fit very well with our model. Therefore we kept or model simple.

In order to model the optical properties of the brain, the measurement of the size and density of cell nuclei in neural tissue were essential. To this end, we looked at specific regions of the zebrafish brain such as the neuropil and periventricular layer (PVL) of the larval zebrafish tectum. We imaged nuclei in those regions using 4',6-diamidino-2-phenylindole or commonly called DAPI staining [6] using a confocal microscope. Figure 2.1b illustrates the difference in neuron density between Neuropil and PVL, which has been quantified and is shown in Fig. 2.1c. In my the simulations, I chose to use the scattering parameters of a region packed with neurons such as PVL.

From Fig. 2.1c, we found that the average radius of a neuron was 4.84, 5.68, and 2.6 μm in X and Y and Z-axes, respectively. We were able to measure the density of neurons in the PVL region and found that this region was packed with neurons at 83.1%.

---

[1]Illumination by unpolarised light is isotropic. If the incoming light is polarised, the scattering is isotropic in a plane perpendicular to the direction of polarization. In both cases, scattering in the forward direction is small.

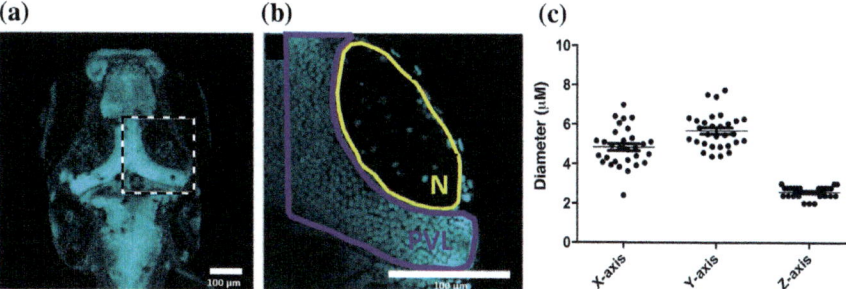

Fig. 2.1 a. Caption of a DAPI staining of neurons of the zebrafish brain. **b.** Zoom in the optic tectum region. Neuropil is marked out in yellow and periventricular layer (PVL) in purple. **c.** Distribution of neuron sizes in PVL

To fully determine the scattering parameters for the model, additional parameters had to be determined such as the average refractive indices of nuclei and its surrounding. This determination and approximation is however very complex. Previous studies suggest that brain tissue without nuclei has an approximate refractive index of 1.34 [7], and nuclei has probably a higher refractive index due to the high refractive index of chromosomes [7–9]. The difference between those two refractive indices is critical for the modelling. To get a refined approximation of those parameters, I have tested different values for the refractive index of nuclei varying from 1.34 to 1.37, with 0.005 steps, and I have calculated the backscattered intensity profiles and compared them with our measurements. The refractive index of 1.35 provided the best fit to the experimental observations, and was therefore used in my modelling. Therefore the refractive index values chosen for the nuclei was 1.35, versus 1.34 for surrounding neural tissue.

The model being fully defined, the next step was to simulate the propagation of a focussed Gaussian beam at a chosen depth and calculate the trajectories of millions of rays (or photons) entering and propagating through the brain tissue. The program initially distributes the rays on a surface with a Gaussian profile. From this initial distribution, it calculates the position and direction of each ray at each scattering event, from the surface until the planes of interests are reached. The four main steps of my calculation are sketched in Fig. 2.2.

The first step of the calculation is to calculate a Gaussian distribution of 4 millions rays leaving the skin in the direction of the focal spot. The next step was to determine the first scattering event position for each ray and its new propagation direction using Monte Carlo method (see next section for mathematical details) taking into account the probability distribution of scattering (see Fig. 2.3).

The next scattering event position was calculated until the focal plane was reached. Once reached, each ray position and direction at the focal plane were recorded and the distribution of positions gives the scattered spot profile at that depth. With the calculation of a sufficient number of rays (depending on the resolution needed), I

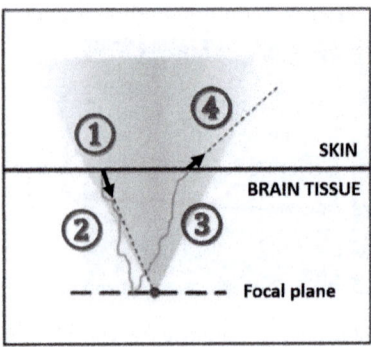

Fig. 2.2 The four steps of the algorithm. 1. Gaussian distribution of (X, Y) positions of 4 million photons on the skin/brain tissue interface (the contributions from skin and above media are ignored). 2. Trajectory calculation in brain tissue for each photon until the focal plane using Monte Carlo method. 3. Backscattered trajectory calculation from the focal plane back to the surface. 4. Calculating of the 4 millions photons final position and direction exiting and reconstructing the image potentially recorded by a camera imaging the focal plane

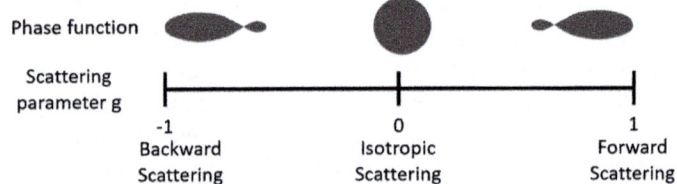

Fig. 2.3 Model of scattering for each scatterer

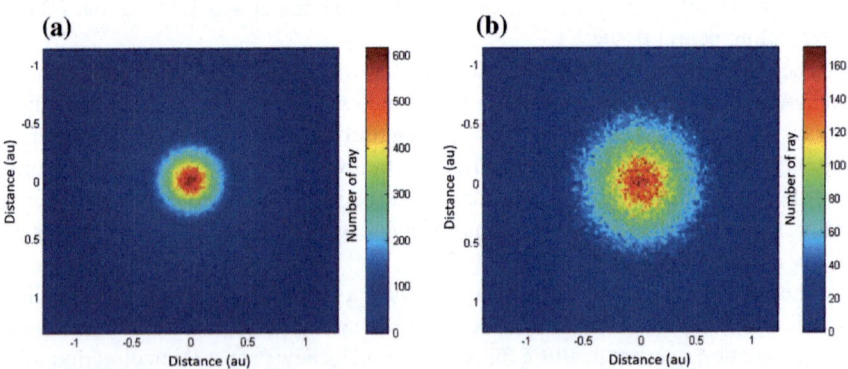

Fig. 2.4 Example of ray distribution for two different scattering parameters with Monte Carlo. Initially 700.000 rays are leaving entering the brain model. The scattering parameters in **a** is smaller than in **b** which results in a broadening of the focussed spot

was able to determine the distribution of light within the brain tissue in the plane of interest (Fig. 2.4 shows an example of ray distribution for two different scattering parameters).

To calculate backscattered light, an additional assumption is required. When a ray reaches the focal plane, it has a certain probability of being transmitted or refracted. However, as the scattering parameter defined by our problem is very close to 1, we can see that the scattering phase function (see Fig. 2.3) shows a similar probability distribution for positive and negative propagation. This means that looking at the forward scattering or backward scattering is nearly equivalent, the main difference being that the event of backward scattering is much less likely than forward scattering event. For time saving reasons, I assumed that every single ray is reflected, and I calculated the trajectories of the rays on the way back to the skin with the Monte Carlo method the same way as previously. Finally, I recorded their positions and directions back at the surface and reconstructed the image expected on the camera. In order to precisely take into account the limited collection angle of the microscope objective in the ray optics simulation, I convolved the Gaussian beam waist of an ideal beam with the result of the Monte Carlo simulation. The details of the mathematical model are presented in Sect. 2.2 and the summary of the results in Sect. 2.3.

2.2 Details of Monte Carlo Method Code Writing for Scattering in Brain Tissue

In this section, I give the mathematical details of the different steps of the Monte Carlo code and model discussed in Sect. 2.1.

I modelled brain tissue as a homogeneous medium with discrete scatterers. The light is modelled as a collection of rays (or photons), where each ray has a given probability per unit distance of encountering a scatterer. The scattering properties of each scatterer are represented by the scattering cross-section and anisotropy parameter [4, 5] which can be calculated using Lorenz–Mie theory [10, 11] for roughly spherical scatterers such as nuclei.

My calculation distributes first a defined number of rays in space. Since I consider the scattering of a focussed Gaussian laser beam, I begin with an initial Gaussian distribution of ray positions (X_0, Y_0, Z_0) as follows:

$$X_0 = normrnd(0, \sigma), \tag{2.1}$$

$$Y_0 = normrnd(0, \sigma), \tag{2.2}$$

$$Z_0 = 0, \tag{2.3}$$

where $normrnd$ is a Matlab function generating random numbers from a normal distribution with a mean of 0 and a standard deviation of σ. The parameter σ is defined by the NA of the microscope objective and the depth at which the Gaussian beam is focussed $Z_{focaldepth}$.

Initially, the rays depart the skin surface towards the focal spot, which gives the initial direction of propagation (V_{X0}, V_{Y0}, V_{Z0}) as follows:

$$V_{X0} = \frac{-X_0}{\sqrt{Z^2_{focaldepth} + X_0{}^2 + Y_0{}^2}}, \tag{2.4}$$

$$V_{Y0} = \frac{-Y_0}{\sqrt{Z^2_{focaldepth} + X_0{}^2 + Y_0{}^2}}, \tag{2.5}$$

$$V_{Z0} = \frac{Z_{focaldepth}}{\sqrt{Z^2_{focaldepth} + X_0{}^2 + Y_0{}^2}}. \tag{2.6}$$

Next, I determine the trajectory of each ray independently, at each distance step. The distance step or distribution of distances, δd, is the distance travelled by rays between scattering events and is expressed as:

$$\delta d = \frac{-\ln R}{C}, \tag{2.7}$$

where R is a function giving a random value between 0 and 1 and C is the cross-section of the spheres.

The probability per unit length P of a ray being scattered is defined as:

$$P = CN, \tag{2.8}$$

where N is the spheres density in the medium.

For a spherical particle, Lorenz-Mie theory provides an analytical solution for the scattering of a plane wave [12, 13]. In the description, the Mie coefficients a_n and b_n are given by:

$$a_n = \frac{\left[\dfrac{D_n(mx)}{m} + \dfrac{n}{x}\right]\psi_n(x) - \psi_{n-1}(x)}{\left[\dfrac{D_n(mx)}{m} + \dfrac{n}{x}\right]\xi_n(x) - \xi_{n-1}(x)}, \tag{2.9}$$

$$b_n = \frac{\left[\dfrac{mD_n(mx)}{m} + \dfrac{n}{x}\right]\psi_n(x) - \psi_{n-1}(x)}{\left[\dfrac{mD_n(mx)}{m} + \dfrac{n}{x}\right]\xi_n(x) - \xi_{n-1}(x)}, \tag{2.10}$$

where ψ_n and ξ_n are the Ricatti-Bessel functions and the logarithm derivative [10, 11]

$$D_n(\rho) = \frac{d}{d\rho}\ln\phi_n(\rho), \tag{2.11}$$

where m is the complex refractive index of the sphere relative to the medium, $x = ka$ is the size parameter, a is the radius of the sphere, and k the wavenumber of incident light. The scattering cross section C_s is then given by [10, 11]:

$$C_s = \frac{2\pi}{k^2} \sum_{n=1}^{\infty} (2n + 1) \left[|a_n|^2 + |b_n|^2 \right]. \tag{2.12}$$

As only scattering by non-absorbing particles is considered, the absorption cross section C_a is zero which means that the total cross section C is equal to the scattering cross section:

$$C = C_s + C_a. \tag{2.13}$$

It is important to emphasize that the medium is treated as homogeneous, the scattering spheres do not have particular positions within the medium. This medium is therefore defined by the probability of scattering per unit length, or equivalently, the mean free path. When a ray is scattered, it will be scattered in a direction dictated by the spheres' distribution of scattering angles. In my model, the distribution of scattering directions is not uniform. Indeed, for the case of low-contrast scatterers of large size (compared to the wavelength), the scattering is predominantly in the forward direction, with very weak backscattering probability. In principle, we could precisely calculate the distribution of scattering angles, $f(\theta, \phi)$, using Lorenz-Mie theory, but using approximation and for simplicity I considered:

$$\theta = \cos^{-1} \left(\frac{1}{2g} \left[1 + g^2 - \left(\frac{1 - g^2}{1 - g + 2gR} \right)^2 \right] \right), \tag{2.14}$$

$$\phi = 2\pi R, \tag{2.15}$$

where g is the anisotropy parameter, and can be expressed using the Mie coefficients as [10]:

$$g = \frac{4\pi}{k^2} C \sum_n \frac{n(n + 2)}{n + 1} \left[\mathcal{R}(a_n a_{n+1}^* + b_n b_{n+1}^*) + \frac{2n + 1}{n(n + 1)} \mathcal{R}(a_n b_n^*) \right]. \tag{2.16}$$

The propagation of each ray can be simulated in the whole region of interest. Once the trajectories of all rays have been calculated, the distribution of rays provides an approximation of the light distribution within the scattering medium.

However, this distribution of rays leaving the surface in the direction of the focal spot would give, in the absence of scattering, an unrealistic zero size focal spot. To correct for this I calculated the convolution of the Monte Carlo ray distribution (once all the ray trajectories have been calculated) in a plane with the focal spot of the unscattered beam calculated using wave theory. For the numerical aperture considered here, the paraxial formula is sufficiently accurate [14], giving a Gaussian focal spot intensity:

$$I(x, y, z) = \left(\frac{w_0}{w(z)}\right)^2 e^{-2\left(\frac{x^2 + y^2}{w(z)^2}\right)}, \tag{2.17}$$

where w_0 is the beam waist defined as:

$$w_0 = \frac{\lambda}{\pi \text{NA}}, \tag{2.18}$$

and:

$$w(z) = w_0 \sqrt{1 + \left(\frac{z}{z_r}\right)^2}, \tag{2.19}$$

where z_r is the Raleigh range, or distance over which the beam radius spread with a factor of $\sqrt{2}$, which can be expressed as:

$$z_r = \frac{\pi w_0^2}{\lambda}. \tag{2.20}$$

After the convolution is calculated, we obtain the distribution of light within the brain tissue.

The next step is to find the backscattered image. Since the backscattering probability is very low, it is time consuming to trace rays until enough of them exit the region of interest in the backward direction. Instead, once the light distribution in the brain tissue was found, I assumed that the backscattering was proportional to the irradiance, and generate a set of backscattered rays. I tracked the rays until they exited the region of interest using the Monte Carlo method. All of the planes above and below the focal spot generate backscattered rays. These will be out of focus, and only light from the focal plane itself will be in focus. They contribute to the overall image captured and needed to be considered in the calculation. If an unscattered Gaussian beam is focussed by the lens, the cross-section of the beam in any given plane is then the point spread function of the lens in that plane. Therefore, I convolved the light from each plane with this point spread function to determine the contribution of that plane to the backscattered image. The last step was to sum up all those contributions to get the predicted camera image of the focal plane in the tissue.

The brain is not a homogeneous and isotropic medium, however, our model considers by approximation a homogeneous and isotropic medium with a certain probability of scattering per distance unit. In addition, the beam is rotationally symmetric and scattering is rotationally symmetric around the beam axis, therefore, averaging the results of a collection of planes around the beam axis gives a more accurate and high resolution light distribution.

2.3 Results of Monte Carlo Method

In Fig. 2.5, I present the intensity contributions of four different planes to the resulting intensity detected by a camera. A total of 4 millions ray trajectories were calculated. From this figure, we can see that the focal plane backscattered rays (in red) contributes predominantly to the central peak of the focal spot observed on a camera, and very little for distances greater than 10 μm away from the focal point centre. The plane positioned at 100 μm deeper than the focal plane (in green) contributes weakly to the total intensity profile. Its flat profile corresponds more to an offset of the total intensity profile. My conclusion from the observations across these depths is that the focal plane contributes mostly to the peak, and distant planes contribute most strongly to offset the intensity profile. Planes near but not at the focal plane contribute to the wings of the intensity profile. The total intensity profile which is the sum of all plane contributions is an infinite sum of Gaussians, where each Gaussian is slightly distorted by scattering in the medium (see Fig. 2.5c). These results apply not only to brain tissue but all scattering media with the same scattering parameters. In addition, the density of scatterers can be varied to simulate different types of neural tissue.

To further understand and predict the effect of scattering on a focal spot, I calculated its dimensions broadening with depth. I first compared a non-scattered and scattered focussed spot at 100 μm deep in tissue in Fig. 2.6. The very first difference is the lower intensity of the scattered spot (in blue) compared to the non-scattered spot (in red) (Fig. 2.6b and c).

As we assumed the light absorption to be 0, we can conclude that in the transverse direction (Fig. 2.6b), scattering spreads the light and in the axial direction (Fig. 2.6c), the beam is not spread but attenuated as the maximum intensity is reduced by the transverse spreading of the focal spot.

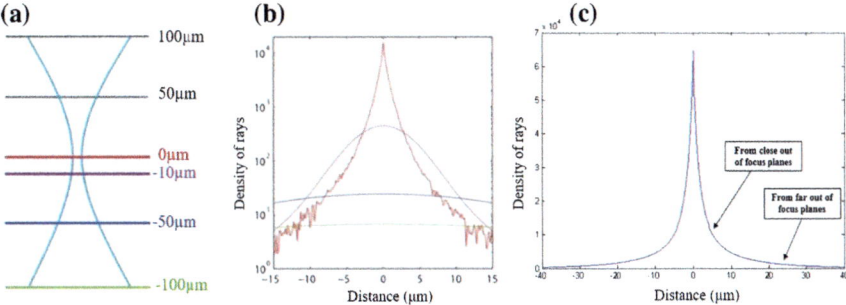

Fig. 2.5 Light contribution to different planes. **a.** A schematic representation of beam shape through 200 μm of tissue, with the focal point at 100 μm depth. **b.** The calculated contributions of intensity and from depths indicated on the left. Red for 0 μm, purple for −10 μm, blue for −50 μm and green for −100 μm. **c.** The result of the summation of all the contributions 100 μm above and below the focal spot located at 100 μm deep with 1 μm steps

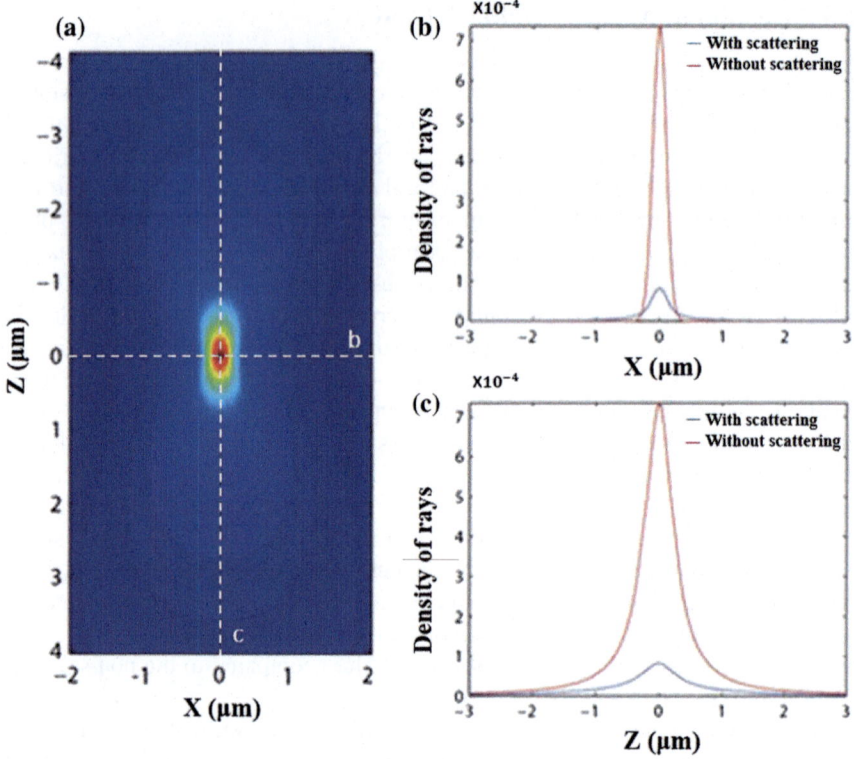

Fig. 2.6 Calculated spreading of a focussed spot. **a.** 2D representation along X and Z axis of the focussed spot after passing through 100 μm of scattering media. **b–c.** Comparison of intensity profile with (blue curve) and without scattering (red curve) along X and Z

In Fig. 2.7 I quantified the spreading and attenuation of light by calculating the full width at half maximum (FWHM) of both spots (solid green and magenta lines in Fig. 2.7a and b).

In the absence of scattering, the size of the focal volume is 0.13 μm (transverse) by 0.29 μm (axial), increasing to 0.16 μm by 0.39 μm at a depth of 100 μm depth. This is an increase of 23% in the transverse direction and 34% in the axial direction. However these values do not give the exact volume increase of illumination as it is in proportion to the laser power. For this reason I also represented the full width at a tenth of the maximum in dashed green and magenta in Fig. 2.7a and b. Finally I performed the calculation of the volume of illumination for two thresholds (10 and 50%) for depth ranging from 0 to 100 μm. I present the results in Fig. 2.7c.

At all depths, the volume illuminated to 50% peak intensity remains relatively small and reaches 0.32 μm for X or Y and 0.78 μm for Z, at 100 μm depth. The volume illuminated to 10% peak intensity is much larger for the same depth: 0.98 μm for X and 3.35 μm for Z.

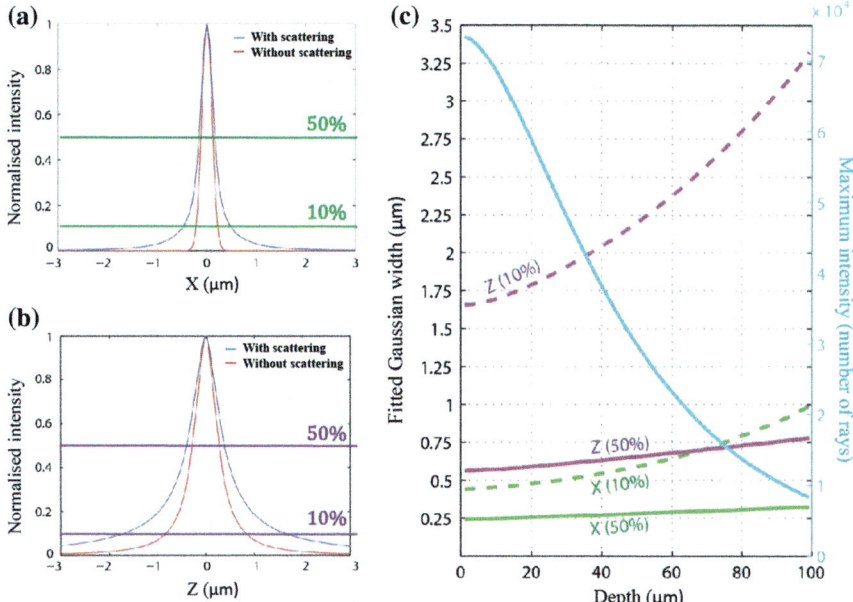

Fig. 2.7 Evolution of a focussed spot dimension within scattering media with depth. **a**. Normalised intensity along X axis for a non scattered (red curve) and scattered (blue curve) beam. **b**. Normalised intensity along Y axis. Purple and green solid and dashed lines represent two different intensity threshold to evaluate the spot size along X and Z axis. **c** Evolution of the spot width with a 10% (dashed lines) or 50% (solid lines) maximum intensity threshold (also represented in a and b) inside brain tissue on the X and Z axes with depth. Evolution of the intensity maximum with depth (light blue curve)

The measured average size of a neuron's nucleus is 4.84, 5.68, and 2.6 μm in X and Y and Z-axes respectively (Fig. 2.1c). From Fig. 2.7c, we can see that a single cell illumination, up to about 70 μm in depth, is possible. Above that depth, the laser power has to be adjusted to limit light spreading to nearby cells.

From the results, I can conclude that the maximum intensity of the illumination volume with depth (blue curve) shows that there is decrease in intensity with depth. At about 40 μm in depth, the initial maximum intensity is halved. This conclusion suggests that when focussing a spot at important depth, the laser power needs to be carefully adjusted to not activate cells on the surface before the focal spot but still activate the neurons deep in tissue.

2.4   Conclusion

The Monte Carlo method developed here was used to simulate the propagation and spreading of a focussed light spot through scattering medium. I was approximating brain tissue to a transparent and very dense brain region. However, this method can be adapted to any brain structure (transparent, non transparent, dense, non dense). As long as the optical properties can be determined, they can be entered in the program developed and all results presented above can be calculated.

In this chapter we were interested in the ability to restrict illumination to single targeted neuron with a focussed laser because we understand that targeting a single neuron or numerous selected neurons in densely packed tissue provides a potentially powerful tool for optogenetic experiments and circuit analysis. As explained above, experiments should be conducted with a careful adjustment of laser power (determined by the light source) and sensitivity (determined by actuator efficiency, expression level, and the target cell's physiological properties) that leads to possible activation/silencing/manipulations only when a large portion of the beam's maximum intensity is targeted on the cell.

In the next chapter, I present experimental measurements in live zebrafish larvae as a means of validating the model presented here.

References

1. I.A. Favre-Bulle, D. Preece, T.A. Nieminen, L.A. Heap, E.K. Scott, H. Rubinsztein-Dunlop, Scattering of sculpted light in intact brain tissue, with implications for optogenetics. Sci. Rep. **5**, 11501 (2015)
2. J.R. Mourant, J.P. Freyer, A.H. Hielscher, A.A. Eick, D. Shen, T.M. Johnson, Mechanisms of light scattering from biological cells relevant to noninvasive optical-tissue diagnostics. Appl. Opt. **37**(16), 3586 (1998)
3. A. Dunn, R. Richards-Kortum, Three-dimensional computation of light scattering from cells. IEEE J. Sel. Top. Quantum Electron. **2**(4), 898 (1996)
4. M.S. Patterson, B.C. Wilson, D.R. Wyman, The propagation of optical radiation in tissue I. Models of radiation transport and their application. Lasers Med. Sci. **6**(2), 155 (1991)
5. S.T. Flock, M.S. Patterson, B.C. Wilson, D.R. Wyman, Monte carlo modeling of light propagation in highly scattering tissues. I. Model predictions and comparison with diffusion theory. IEEE Trans. Biomed. Eng. **36**(12), 1162 (1989)
6. J. Kapuscinski, Dapi: a DNA-specific fluorescent probe. Biotech. Histochem. **70**(5), 220 (1995)
7. R. Barer, Refractometry and interferometry of living cells. J. Opt. Soc. Am. **47**(6), 545 (1957)
8. J.L. Sandell, T.C. Zhu, A review of in-vivo optical properties of human tissues and its impact on PDT. J. Biophotonics **4**(11–12), 773 (2011)
9. B. Rappaz, P. Marquet, E. Cuche, Y. Emery, C. Depeursinge, P.J. Magistretti, Measurement of the integral refractive index and dynamic cell morphometry of living cells with digital holographic microscopy. Opt. Express **13**(23), 9361 (2005)
10. C.F. Bohren, D.R. Huffman. *Absorption and Scattering of Light by Small Particles*
11. H.C. Van de Hulst, Light scattering by small particles. Q. J. Royal Meteorol. Soc. **84**(360), 198 (1958)
12. L. Lorenz, Lysbevaegelsen i og uden for en af plane lysbolger belyst kugle. Vidensk. Selsk. Skr. **6**, 2 (1890)
13. G. Mie, Beitrage zur Optik truber Medien, speziell kolloidaler Metallosungen. Ann Phys. **330**(3), 377 (1908)
14. T. Nieminen, H. Rubinsztein-Dunlop, N.R. Heckenberg, Multipole expansion of strongly focussed laser beams. J. Quant. Spectro. Radiat. Transf. **79–80**, 1005 (2003)

Chapter 3
Scattering in Zebrafish Brain for Optogenetics

In the previous chapter, I focussed on the theory and modelling of light scattering in brain tissue. In this chapter, I present the measurements of backscattered light in-vivo in the zebrafish brain and compare those with my Monte Carlo method and model. I measured backscattered light in areas rich and poor in cell bodies, and compared them to identify the distinct and dramatic contributions that cell nuclei make to scattering. These results have been summarised and published in the following paper:

I. A. Favre-Bulle, D. Preece, T. A. Nieminen, L. A. Heap, E. K. Scott, and H. Rubinsztein-Dunlop. *Scattering of sculpted light in intact brain tissue, with implications for optogenetics.* Scientific Reports **5**, 11501 (2015).

3.1 Description of Set-Up to Study Light Scattering in Brain Tissue

Ideally, to be able to precisely measure how a spot of light broadens with depth, or brain tissue density, we would need to measure the light intensity at the focussed spot plane in tissue directly, without distortions that light encounters on the way back to the camera. However, this seems impossible as tissue surrounds the focussed spot of light. I have considered several possible solutions:

- The dissection of brain tissue changes its optical properties within minutes, therefore this is not a feasible solution.
- The introduction of fluorescent particles in the region of the focal spot, acting as point sources, could help us define the optical properties of the brain. However, it would be extremely hard, if not impossible, to arrange them in one single plane, to measure the intensity received in discrete positions away from the focussed spot.

© Springer International Publishing AG, part of Springer Nature 2018
I. A. Favre-Bulle, *Imaging, Manipulation and Optogenetics in Zebrafish*,
Springer Theses, https://doi.org/10.1007/978-3-319-96250-4_3

- Measuring the light transmission through the sample, instead of the backscattered light, could be a more direct measurement of light distribution, however, the light would then have to pass through the fish jaws and gills, a highly scattering area.

For all those reasons, I decided to look at backscattered light with the same objective that is used to create a focussed spot. The Monte Carlo code allows the calculation of light scattering on the way in and out, which will allow us to compare results from calculations and measurements.

To address the problem of light scattering, and its measurement through-out brain tissue, an optical system was set up using the combination of a laser and a spatial light modulator (SLM described below). This system allows the focus of diffraction-limited spots of light within different brain regions with different scattering parameters. Using this system, we were able to measure the backscattered light from a single diffraction limited spot.

3.1.1   Introduction to Spatial Light Modulators (SLM)

A Spatial Light Modulator (SLM), as its name indicates, can modulate light spatially. SLM modulates the light in one plane (the screen plane) and its pattern displayed can be modified to position or deliver light very precisely in space. An SLM is basically a display where each pixel is composed of liquid crystal and electrodes changing the optical properties of the liquid crystal (Fig. 3.1).

The electrodes create a potential difference in the liquid crystal which orientates the molecules of this liquid crystal. The orientation of those molecules causes the refractive index, perceived by the reflected beam on a pixel, to change. This is equivalent to introducing a phase shift (or time delay) to the output beam.

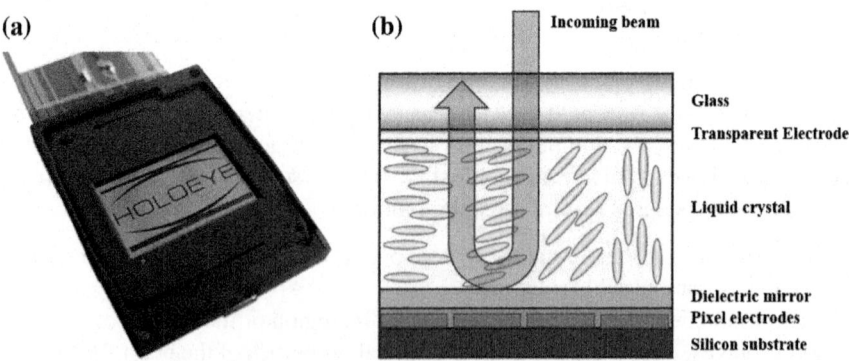

Fig. 3.1 **a** Example of a Spatial Light Modulator (SLM) screen. **b** Sketch of SLM pixel structure. The light reflected from a pixel goes through liquid crystal with an orientation dictated by the voltage applied on the electrodes. This technique allow the control of the outgoing beam time delay

A high quality SLM, which is a function of the depth of the cell, can introduce 2π or more phase shift which means that they have a total control of the beam wavefront and can modulate it as precisely as the resolution of the SLM screen can be.

To illustrate further this point, I present in Fig. 3.2 the equivalence of a prism and a lens in terms of phase delay that are applied to SLM. A prism has the property of deviating a beam, or changing the angle of an incoming beam, which results in a displacement in 2D (X and Y) after transmission trough a lens or microscope objective. A lens has an ability to change the position of a point in one dimension (Z). With the combination of lenses and prisms, we can move one or multiple spots in three dimensions by simply adding their phase delays. The information sent to the SLM is a grey scale image, where each pixel value represents the phase delay we want to introduce to the beam at that pixel position. This grey scale image is called hologram. Using this method, we can calculate a hologram that we display on the SLM screen and alter a beam wavefront to move a spot in 3 dimensions.

Of course SLMs do not only create spots in 3D, but can, with the use of algorithms, shape the light to arbitrary patterns in two dimensions [2, 3]. Shaping light in three dimensions is also possible but for simple or specific geometries [4–6]. The light distribution within the 3 dimensional shaping however remains uneven and is not robust enough for in-vivo illumination.

SLMs, however, have some light efficiency limitations:

- As their screen is made out of pixels, the spacing between its pixels introduces light loss. Typically, the fill factor, or fraction of the effective area over the array of pixels area, is between 80 to 90%.
- The relationship between phase delay on the reflected beam and voltage at the pixel is not linear. Several experimental techniques allow to find out its look up table (LUT) for any wavelength used [7].
- The more deflection we want to apply to the light, the less efficient an SLM is. Light intensity typically evolves as a Sinc function of distance from the optical axis.

(a)   **(b)**

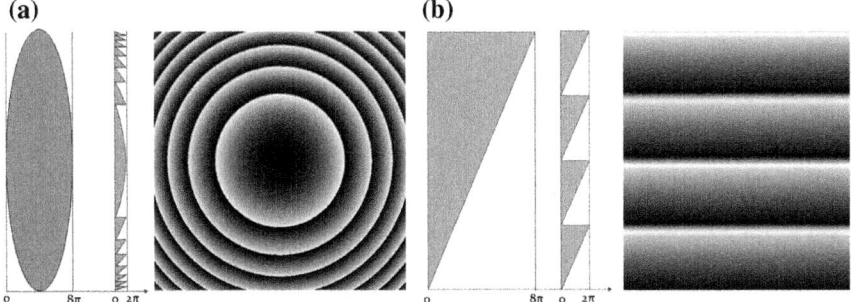

Fig. 3.2 a Example of a prism hologram: Calculation of the phase modulation 2π for a prism gives triangular steps. **b** Example of a lens hologram based on the same principle

- The un-deflected light is called the zero order and must be removed for some applications as this light interferes with the measurement. Spatial filters such as irises are commonly used to reduce unwanted light contamination.
- The necessity of phase discontinuity, in value (bits per pixel) and spatially (pixels are discrete and have a specific pitch), in SLMs results in the appearance of higher orders, also called ghost orders. Those multiple orders can dramatically affect the illumination depending on its geometry, however, pre-existing algorithm have been developed to practically remove them [8–10].

Despite those limitations, a SLM's flexibility and wavefront modulation performance is very attractive. This has very often led researchers to use SLMs in very complex media or to combine them with additional optical systems.

For example, the use of SLM led to the possibility of multi-neuronal activations and multi-plane imaging over large volumes of brain tissue [2, 5, 11–13]. The possibility to illuminate multiple single cells deep in-vivo, and relate their activity to the

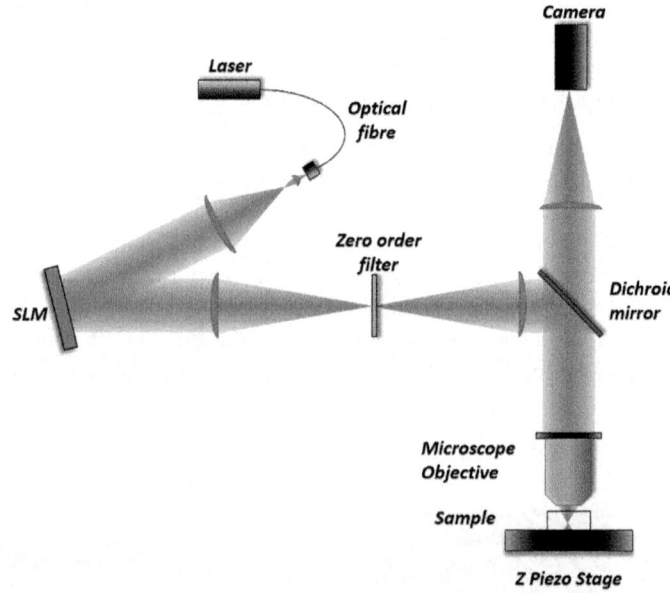

Fig. 3.3 Experimental set-up for studies of scattering. For illumination, we used a 150 mW, 488 nm OBIS laser coupled to an optical fiber (single mode for visible light). The laser light was expanded and collimated to fill the screen of a Spatial Light Modulator (SLM PLUTO VIS Holoeye). The SLM screen plane was imaged at the back focal plane of a 40× water immersion microscope objective (Olympus LUMPLFLN 40XW, 0.8 NA), slightly overfilling the back aperture to guarantee a diffraction limited spot at the focus. The SLM's total light efficiency for the generation of a single spot is about 60% for visible light. To remove the non-diffracted light of an SLM (or zero order), we used a spatial filter in the optical path (see next section for details). Images were acquired by collecting the reflected light of a single spot through the same microscope objective. The reflected light passed through a dichroic mirror (reflecting 488 ± 10 nm and transmitting other visible light) and was directed to a CMOS camera (PCO Edge) after passing through a single lens

rest of the brain, is essential for the study of brain circuitry. It allows to conclude on the entire processing of the information and also to determine if specific neurons are sufficient or essential to the circuitry.

3.1.2 Experimental Set-Up for Studies of Scattering

The apparatus is presented and detailed in Fig. 3.3. The flexibility of SLM allows us to focus multiple diffraction limited spots in 3D in brain tissue.

An SLM displays a grey scale images which are calculated from the image we want to achieve at the microscope objective imaging plane. From the above set-up configuration, we can see that the SLM screen is imaged at the back-focal plane of the microscope objective, which means that the illumination at the focal plane of the microscope objective is the Fourier transform of the image displayed on the SLM (and vice-versa).

Programs which interactively calculate holograms exist, they have been developed in the optical trapping community and are available online. The one we used in this project is from the Padgett group at the School of Physics and Astronomy at the University of Glasgow:

http://www.gla.ac.uk/schools/physics/research/groups/optics/research/optical tweezers/software/

This program creates as many spots as wanted in 3D. I implemented this program to our system for an easy manipulation of multiple spot in 3D. An example of hologram calculated with this program, and the corresponding image at the focal plane of microscope objective, is presented in Fig. 3.4.

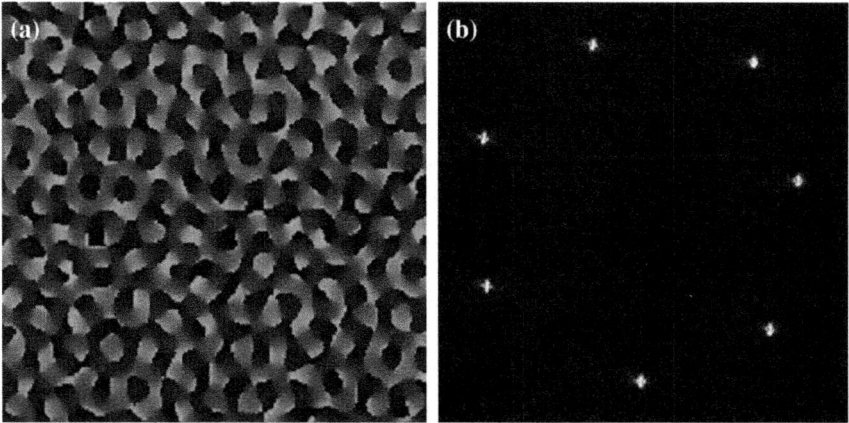

Fig. 3.4 Example of hologram displayed on an SLM in (**a**) and corresponding image in the Fourier plane in (**b**) In this case the aim was to create 7 spots equally spaced along a ring. The left figure demonstrates the pattern used to obtain the configuration of 7 spots

3.2   Experimental Measurements

3.2.1   Experimental Procedures

To measure the scattering of light in neural tissue in zebrafish in-vivo, I need to immobilize them with minimal disturbance of their environment. Zebrafish are kept in E3 medium which is a common medium for embryos. The composition of the medium is salt water and methylene to prevent fungal growth.

To immobilise them, one can use low melting agarose (LMA). This agarose sets at ambient temperature and immobilizes the larvae but still allows it to breathe. By mixing E3 and LMA at different concentrations, I found that 2% LMA is a good compromise between, a too weak embedding where embryos can escape, and a too strong one where its body could feel squeezed. Moreover, larvae zebrafish can stay for hours in 2% LMA without being affected.

A 6dpf larvae with LMA was set up on a microscope slide and positioned under the microscope objective at the correct height with a Z piezoelectric stage and manual (X, Y) stage. Finally, a diffraction-limited focussed spot was shone into the brain region of interest and the backscattered light was recorded on the camera.

3.2.2   Measurements

An example of backscattered intensity profile is presented in Fig. 3.5a. It has a consistent shape all over the brain but with a width and intensity varying depending on depth of focus and region of the brain. I verified that this intensity profile could be nicely fitted with a sum of two Gaussians as shown in Fig. 3.5b. Each Gaussian can be expressed as follows:

$$f(x) = \frac{1}{\sigma\sqrt{2\pi}} e^{\left(\frac{-x^2}{2\sigma^2}\right)}, \tag{3.1}$$

where the parameter σ is related to the full width at half maximum (FWHM). By approximation, in the following results I considered the width of each Gaussian to be 4σ.

Quantifying scattering here is equivalent to measuring the broadening of a diffraction limited spot in terms of width, but also peak intensity attenuation. Figure 3.6a and c show the width of the two Gaussians with depth in the Neuropil and PVL. We can see that at the skin surface (0 μm in depth) the two Gaussians have the same width which is the width of the diffraction limited spot. In both cases when going deeper into brain tissue the wider Gaussian width (in red) increases with depth while the narrower Gaussian width (in blue) has a relatively consistent value of few microns with depth. This result suggests that the blue Gaussian represents the original diffraction limited spot attenuated and the red Gaussian the scattered photons which create the blurriness of the spot.

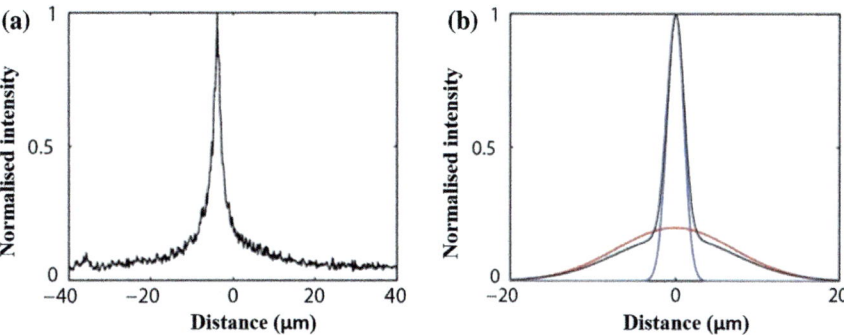

Fig. 3.5 Example of measurements of backscattered light. **a** Example of a backscattered intensity profile recorded from a spot focussed $100\,\mu$m deep in brain tissue. **b** The profile presented in a can be approximated to be a sum of two Gaussian. The blue Gaussian (the narrower) represents highest intensity component. The red Gaussian (the wider one) represents the wings of the intensity profile. (The black line represent their sum)

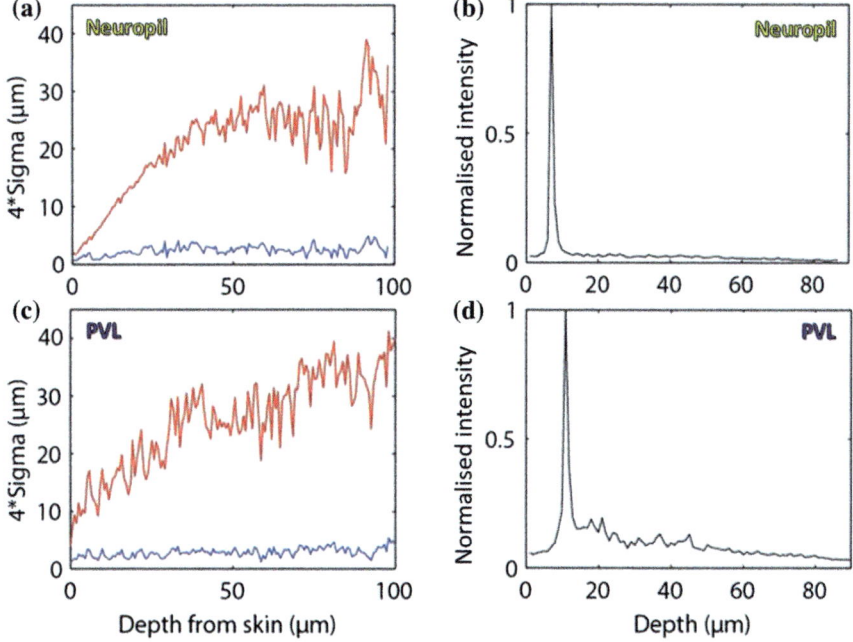

Fig. 3.6 The evolution of the fitted Gaussian width with depth is shown for the tectal neuropil (**a**) and PVL (**c**), where the narrow and intense Gaussian is represented in blue, and the wider and weak Gaussian is represented in red. **b** and **d** show measured backscattered intensity maxima with depth in neuropil and PVL, respectively

When looking at the maximum of intensity backscattered with depth (Fig. 3.6b and d), surprisingly, a high peak of reflectivity was observed at about 10 μm. This peak corresponds to the larva's skin. The mismatch of refractive index between the skin and LMA becomes obvious. As the focal spot is moved deeper, the contribution of the skin reflection decreases, and we can observe the predominant effect of scattering within the brain. The main difference between neuropil and PVL regions is that after this high peak of intensity due to the skin, in the PVL region, we find a relatively high fluctuation of maximum intensity compared to the neuropil region where the backscattered light is nearly zero. The neuropil region, being very poor in neurons compared to PVL region, theoretically would not backscatter much light as the refractive indexes mismatches are not as strong as the PVL region. These assumptions are confirmed as we can see that the presence of neuronal bodies increase dramatically the amount of backscattered light of scattered light in general.

3.3   Comparing Monte Carlo Method with Experimental Measurements

In Chap. 2, I presented my simplified model of brain tissue and the use of Monte Carlo method to calculate the trajectory of rays (of photons) within the tissue depending on its scattering properties. I summed all contributions from 201 planes, with 1 μm intervals, from 100 to −100 μm depth relative to the focus to provide an accurate estimate of what should be recorded on the camera and I found that the final recorded intensity profile is more of an infinite sum of Gaussians than just the sum of two. In Fig. 3.7a, I overlaid the experimental measurement and my model of the image recorded at 100 μm depth.

My simulation (blue curve) matches very well the experimental measurement (black curve). The difference is an offset present in the experimental measurement, which is greater than in the simulation. This can be explained by higher reflectivity at the skin, which we observed in Fig. 3.6 panels b and d. It could also be explained by the presence of small scatterers (which tend to scatters at high angles) not considered in the model.

This convincing fitting shows that the actual intensity profile is not a simple sum of two Gaussians, as presiously assumed, but the sum of a theoretically infinite number of Gaussians, each Gaussian deformed to a different degree by scattering.

Further investigation of the model was directed toward a calculation of the decrease in the maximum peak intensity of the backscattered light with depth. In Fig. 3.7b, I represented in blue the result of my Monte Carlo code, however, to take into account the highly reflective skin, I considered independently the contribution of a simple flat reflective surface (represented in red), and its sum (in black) fits relatively well the experimental measurements (Fig. 3.7c). The sawtooth nature of the experimental curve can be understood as the effect of nuclei, which serve as points of high backscattering. In contrast, the model produces a smooth curve, since

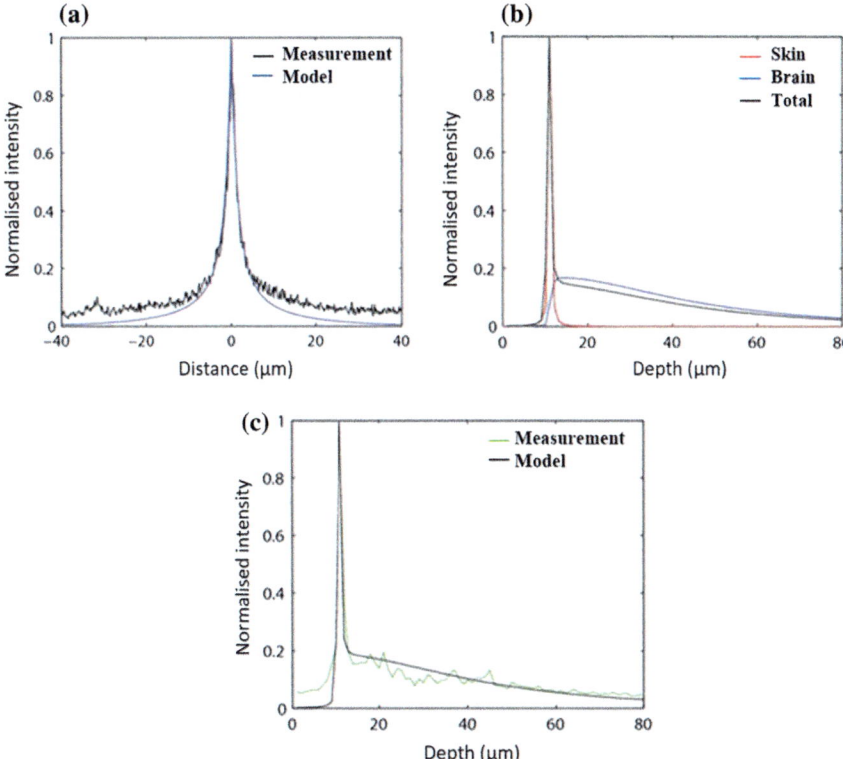

Fig. 3.7 Backscattered measurements with depth and comparison with Modelling. **a** Sum of all calculated contributions from 201 planes (blue) registered against experimental observations (black). **b** Calculated backscattering from skin (red) and brain (blue), and total calculated backscattering (black), by depth. **c** total calculated backscattering (black, drawn from b) registered against experimental measurements in the PVL (green, drawn from Fig. 3.6d)

scattering is simulated in an homogeneous scattering medium, and therefore cannot be punctuated spatially.

Finally we can see that the calculated and observed backscattering profiles and their evolution with depth suggest that our calculated distribution of light is essentially correct and the theoretical results presented in Chap. 2 are validated.

3.4 Conclusions

The intensity profiles recorded on the camera do not represent the spatial distribution of light within brain tissue at the focal spot, which is what we are after as the actual distribution of light is critical for optogenetic manipulations. However, what I show here is by constructing a simplified model of brain tissue and scattering using the

Monte Carlo method, the calculation of distribution of light within the brain tissue can be deduced from the calculation of backscattered light measured on the camera, as its calculation has been done few steps ahead of the final image calculation (Fig. 2.2).

My theoretical model of light scattering, being validated by the experimental measurements, demonstrates that volumes equivalent to single neurons can readily be illuminated at relatively important depths in neural tissue.

Furthermore, these results demonstrate that, through the use of an SLM, multiple volumes, equivalent to single neurons, can readily be illuminated at depth in intact neural tissue with careful consideration of the total illumination power. Also, the use of 488 nm laser (low wavelength), suggests that this targeting performance can be improved with the use of longer wavelength lasers, as the tissue would appear clearer at those wavelengths.

Interestingly, restricting illumination to single targeted neurons, or to a number of selected neurons in densely packed tissue, is very attractive as it would provide a potentially powerful tool for circuit analysis. Indeed, as SLMs can be used to project numerous spots of light in three dimensions, they could be used to drive, or silence, neuronal patterns of activity throughout a small brain, which would lead to a better understanding of neuronal circuitry and behaviour.

In Chap. 4, based on the results presented in this chapter, we build an optogenetic system for delivering such sculpted light into the brains of zebrafish.

References

1. I.A. Favre-Bulle, D. Preece, T.A. Nieminen, L.A. Heap, E.K. Scott, H. Rubinsztein-Dunlop, Scattering of sculpted light in intact brain tissue, with implications for optogenetics. Sci. Rep. **5**, 11501 (2015)
2. A.M. Packer, B. Roska, M. Hausser, Targeting neurons and photons for optogenetics. Nat. Neurosci. **16**(7), 805 (2013)
3. E. Papagiakoumou, F. Anselmi, A. Begue, V. de Sars, J. Gluckstad, E.Y. Isacoff, V. Emiliani, Scanless two-photon excitation of channelrhodopsin-2. Nat. Meth. **7**(10), 848 (2010)
4. W. Graeme, C. Johannes, Experimental demonstration of holographic three-dimensional light shaping using a gerchberg-saxton algorithm. New J. Phys. **7**(1), 117 (2005)
5. S. Yang, E. Papagiakoumou, M. Guillon, V. de Sars, C.M. Tang, V. Emiliani, Three-dimensional holographic photostimulation of the dendritic arbor. J. Neural Eng. **8**(4), 046002 (2011)
6. V.R. Daria, C. Stricker, R. Bowman, S. Redman, H.-A. Bachor, Arbitrary multisite two-photon excitation in four dimensions. Appl. Phys. Lett. **95**(9), 093701 (2009)
7. D. Engstrom, M. Persson, J. Bengtsson, M. Goksor, Calibration of spatial light modulators suffering from spatially varying phase response. Opt. Express **21**(13), 16086 (2013)
8. R. Bowman, V. D'Ambrosio, E. Rubino, O. Jedrkiewicz, P. Di Trapani, M.J. Padgett, Optimisation of a low cost slm for diffraction efficiency and ghost order suppression. Eur. Phys. J. Spec. Top. **199**(1), 149 (2011)
9. E. Ronzitti, M. Guillon, V. de Sars, V. Emiliani, Lcos nematic slm characterization and modeling for diffraction efficiency optimization, zero and ghost orders suppression. Opt. Express **20**(16) (2012)
10. M. Agour, E. Kolenovic, C. Falldorf, C. von Kopylow, Suppression of higher diffraction orders and intensity improvement of optically reconstructed holograms from a spatial light modulator **11**, 105405 (2009)

11. A.M. Packer, L.E. Russell, H.W.P. Dalgleish, M. Hausser, Simultaneous all-optical manipulation and recording of neural circuit activity with cellular resolution in vivo. Nat. Meth. **12**(2), 140 (2015)
12. W. Yang, J.E. Miller, L. Carrillo-Reid, E. Pnevmatikakis, L. Paninski, R. Yuste, D.S. Peterka, Simultaneous multi-plane imaging of neural circuits. Neuron **89**(2), 269 (2016)
13. S. Quirin, J. Jackson, D.S. Peterka, R. Yuste, Simultaneous imaging of neural activity in three dimensions. Front. Neural Circuits **8**, 29 (2014)

Chapter 4
Optical Systems to Decode Brain Activity

Different methods have been developed over the past decades for studies of brain function. These methods aim to decode brain activity in terms of the communication that takes place among neurons, and to identify the patterns of neural activity that ultimately produce behaviour.

By delivering precise stimulation to one brain region, and tracing the resulting neural activity, one can build a map of how information flows through the brain's circuits. This is conceptually simple, but its application requires sophisticated optical physics and microscopy. In this chapter, I describe my construction of a system capable of such functional brain mapping.

4.1 Fluorescent Proteins and the Development of GECIs

Neurons communicate by transmitting electrical signals trough their axons which are transformed into chemical signals (neurotransmitters) at their synapses to the next neuron (Fig. 4.1).

Electrophysiology is a common technique for studying electrical signals of nervous or other bodily activity [1–4]. It tests and measures electrical activity on a wide variety of scales, from single cell to an entire organ. It uses an electrode, placed in the tissue, and measures the flow of ions which gives the measure of the activity of that region. With this technology, one can detect and measure single cells activity. However, to study single neuron activity, this technique is quite invasive as the cell needs to be reached by the electrode.

As we want to fully understand the wiring conditions and investigate which neurons are necessary or sufficient for the information processing in response to a certain stimuli, we need to address each neuron individually, or address a restricted number of neurons. One method widely used now to permanently silence single neurons

© Springer International Publishing AG, part of Springer Nature 2018   33
I. A. Favre-Bulle, *Imaging, Manipulation and Optogenetics in Zebrafish*,
Springer Theses, https://doi.org/10.1007/978-3-319-96250-4_4

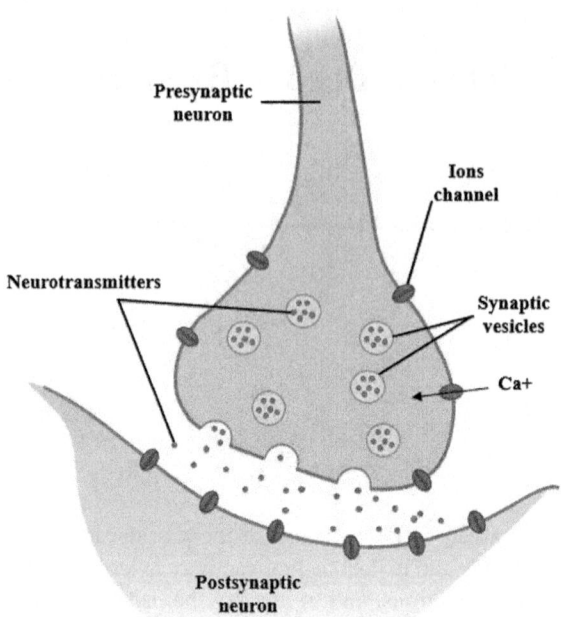

Fig. 4.1 Sketch of processes in neuronal synapses. An electrical signal is transmitted through the pre-synaptic neuron to reach the synapse. This affects the ion channels and induces ions transfers which induces the formation on synaptic vesicles and the release of neurotransmitters at the synapse detected by the post-synaptic region of the following neuron

is their destruction with highly intense 2P lasers exposures [5, 6]. Once more, this method is very invasive but also irreversible.

Recently, researchers found that cells can be genetically modified to express light sensitive ion channels. Those channels are located on the cells surface and allow the transfer of ions, triggering or inhibiting activity. Those proteins are called opsins and allow light to control the activity of cells in living tissue. This discovery led to the development of optogenetics, a non-invasive technique able to silence or trigger activity in cells using light [7, 8].

In optogenetics, we firstly want to express a protein in the cells of interest to sensitize them to light, and secondly to develop an optical system to deliver light to those specific cells very precisely in order to control them. Furthermore, in order to gauge the function of those cells, we need to visualize their activity and its spreading. This requires a parallel optical system for imaging GECIs.

Opsins are a group of light-sensitised proteins that are commonly use in optogenetics. Two widely used opsins are Channelrhodopsin (ChR2) [9, 10] and Halorhodopsin (NpHR) [11, 12] (Fig. 4.2). Channelrhodopsin channels positive ions only (Na^+) from the outside to the inside of the cell which increases the cell's potential and triggers an electrical signal. In contrast, Halorhodopsin channels only negative ions (Cl^-) which inhibits the signal.

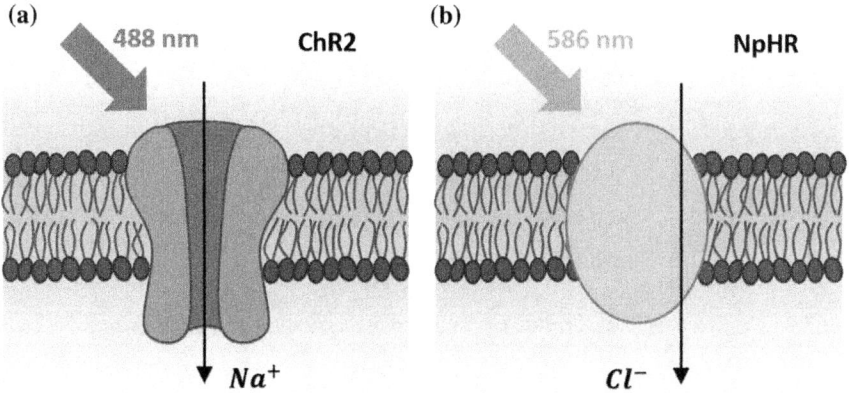

Fig. 4.2 Channelrhodopsin (ChR2) and Halorhodopsin (NpHR) light sensitive ion channels. **a** ChR2 channels Na^+ ions into cells when illuminated by 488 nm light increasing its potential resulting in the transmission of an electrical signal. **b** NpHR channels Cl^- ions inside cells when illuminated by 586 nm light decreasing its potential resulting in the inhibition of its activity

Depending on the study, and questions one wants to answer, one protein or the other should be expressed. Once the cells are light sensitized, their intense illumination can trigger or inhibit activity.

To visualise this activity and its spreading throughout the brain, we need additional techniques. Recently, calcium imaging has emerged as an approach for overcoming some of these liabilities. Because the concentration of calcium ions rises when a neuron is active, genetically-encoded calcium indicators (GECIs) [13, 14], which become more brightly fluorescent when calcium is high, report indirectly on a neuron's electrical activity. This technique is much slower and less sensitive [15] than electrophysiology, but comes with the great advantage that it can report on thousands of cells simultaneously [16–19]. The challenge lies in developing microscopes capable of this imaging, and in using a model system well suited to such optophysiology.

4.2 Imaging Neuronal Activity

With the existing GECIs and opsins, one can express those proteins in specific neurons or brain regions. This will allow to control those neurons and to visualize them. The optical systems for activity control and imaging have to work in parallel to guarantee a precise control and imaging to permit functional circuits to be mapped.

However, light delivery into brain tissue is a very complex problem as absorption and scattering occur and these effects are difficult to predict in complex tissues. A cell is highly heterogeneous in terms of refractive index as organelles present in the cell are small high refractive index structures in a low refractive index medium (close to water). At the same time the nucleus of a cell also has a very high but a heterogeneous refractive index. Keeping this in mind, researchers have built and are still designing optical systems able to deliver light deep inside tissue with minimal distortions [20–23].

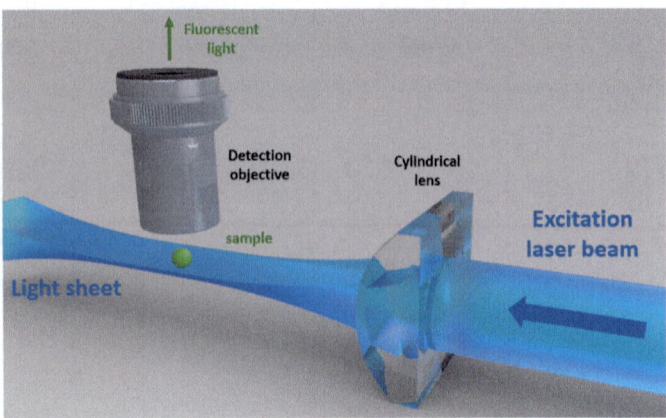

Fig. 4.3 Sketch of a common configuration of a Selective Planar Illumination Microscope (SPIM). A light sheet is created by a cylindrical lens and focussed perpendicular to an imaging microscope objective

In this section, we describe some optical techniques currently used in optogenetics to image neuronal activity with minimal light distortion and therefore higher imaging resolution. We will discuss some optical tools and their combination to obtain further precision of light delivery.

One simple optical method for light delivery of a specific wavelength is optical fibres [24–26]. Optical fibres have the advantage of working with non transparent brain (like mice) but are invasive as they have short working distances. New techniques involve the coupling of spatial light modulators (SLM) and optical fibres [27, 28] for longer working distances and more complex light shaping.

To overcome invasive methods, Selective Planar Illumination Microscopy (SPIM) has been developed for transparent samples [29–32]. A SPIM creates a light sheet and delivers light in only one highly defined plane (Fig. 4.3) which limits light noise from planes above and below the imaging plane and limits the sample light exposure. A SPIM is usually positioned perpendicular to the imaging objective so that the whole plane of light is imaged on a camera.

A SPIM is commonly constructed using a cylindrical lens which focusses light in one dimension and results in a thinner light sheet when transmitted by a microscope objective. A detailed SPIM is presented in Fig. 4.4.

The advantage of having a single lens instead of a microscope objective [33, 34] is a long working distance (few centimetres) which allows room for a chamber holding the sample. Also the length of the light sheet is about the same as the incoming beam size which is much larger than the sample and very convenient for imaging. The thickness D of a light sheet depends on the numerical aperture (NA) of the cylindrical lens and the wavelength λ and their relationship is as follows:

$$D = 1.22\frac{\lambda}{NA}. \tag{4.1}$$

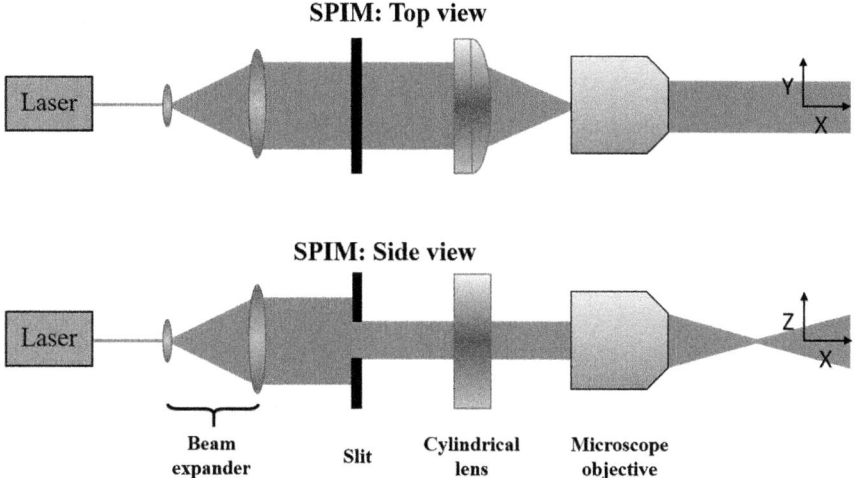

Fig. 4.4 Optical configuration for SPIM. A laser beam goes through two lenses in a telescope configuration to increase the beam size, a horizontal slit, a cylindrical lens and finally a microscope objective imaging the slit into the same

The higher the NA of the lens, the thinner the light sheet is. However for single cylindrical lenses, the NA is very low which creates planes much larger than a single neuron (around 5 μm). This leads us to use microscope objectives that have higher NA in order to create thin light sheets. However, its length L, in the beam propagation direction is defined as follows:

$$L = 1.67 \frac{\lambda}{NA^2}. \tag{4.2}$$

A compromise has to be made between the thickness of the plane and the length of the light sheet. There are now other methods being developed to provide larger length of the sheet of light. These are based on Airy beams that can be created using SLMs [35].

SPIM has many advantages to work with and it is now widely used as it allows low phototoxicity [36], high speed and volumetric imaging [37–39].

Another optical system widely used in optogenetics is SLM (see Sect. 3.2.1 for details). They modify light wavefronts and help delivering light very precisely in space which enables a high flexibility in the optical system. Past studies have shown that they can help targeting a chosen volume of cells, numerous single neurons independently and simultaneously [40], or can target very precise portions of axons [41].

This non exhaustive presentation of optical tools for optogenetics shows how versatile they can be depending on the application or system to be studied. In many cases, including the ones described in this thesis, experiments require that the correct combination of optical approaches be delivered on a single microscope.

4.3   The Advantages of Combining Optical Systems

The flexibility of SLM has led researchers to combine them with additional optical systems. An interesting example of SLM use is the correction of distortions and aberrations introduced by the sample media or the optical system itself [23, 42, 43]. These possibilities make it very attractive for any imaging or illumination system.

Another interesting example of light delivery technique with SLM involves the coupling of SLM and 2-photon excitation laser to decrease the illumination spreading in Z of a 2D illumination. To further control illumination volumes, spatio-temporal shaping of lasers pulses can be implemented [44]. I will discuss further light shaping for single photon excitation laser in the next chapter when I implement the Gerchberg-Saxton algorithm to shape light in 2D.

A new technique called Swept Confocally-Aligned Planar Excitation (SCAPE) has been recently published and combines an oblique SPIM with an oblique imaging system on the same microscope objective. The alignment on the SPIM illumination and the imaging detection is such that a single galvo mirror deflects the illumination and the detection identically, which allows very fast and deep 3D scans in transparent samples. However such a system is difficult to implement and the fact that the imaging and detection planes are non perpendicular requires data post-processing which further complicates the interpretation of results.

Imaging techniques and their combinations are numerous and all the combinations of optical systems presented here are useful or essential for some applications. Decoding brain activity is a real challenge and the correct combination of optical systems for the problems we want to solve is very challenging. Understanding the advantages and limits of those different optical techniques and systems for our application is essential.

4.4   Building of the Optogenetics System

As outlined above, the new-found abilities to manipulate targeted neurons, and to observe the resulting circuit activity throughout the brain, open up unprecedented opportunities for mapping the brain's function. This approach requires multiple sophisticated optical systems to operate in parallel, and as a result, the implementation is complex. In this chapter, I describe the custom microscope and illumination apparatus that I have designed and built, and provide an example of a brain-mapping experiment in which it has been used. In this case, the goal was to illuminate a region of the brain (the thalamus), rather than individual neurons, so I have further developed a method for creating large custom holograms using an SLM.

4.4.1 Presentation of the Optogenetics System

After looking into the literature to find the ideal optical system to image and drive brain activity in zebrafish, we made multiple choices based on the advantages and drawbacks of each one which led us to our own combination of genetics and optical systems:

- GCamp6f protein for calcium imaging [15, 45]. It is the fastest genetically-encoded calcium indicator (for cytoplasmic free calcium) in neurons which makes it very attractive for fast acquisitions. Its excitation wavelength is 488 nm and emission 512 nm.
- Channelrhodopsin (ChR2) protein for cell light sensitizing [9, 10]. Its activation wavelength is 470 nm and is expressed in the whole brain in the context of this thesis. We can notice that the excitation wavelength for ChR2 and GCamp6f are very similar and that a 488 nm laser with high enough power can trigger both proteins. However we will see latter in this chapter that for low enough power of 488 nm laser, we can visualise brain cells without triggering activity. Indeed, as the ChR2 peak is at 470 nm, it needs a stronger power to have any effect on cells behaviours.
- SPIM for brain activity imaging [29, 30, 46]. The transparency of the brain and its relative small size allow SPIMs to be not too much distorted. Very recent microscope objectives have the characteristics of having a decently large numerical aperture for a very long working distance such as the OLYMPUS XLFLUOR 4X/340 Macro objective with an NA of 0.28, and a working distance of 29.5 mm. With such objectives we can create light sheets of about 2.13 μm along Z axis and 1–2 cm along Y axis (Fig. 4.4).
- SLM for light sensitive cells targeting. Once again, the transparency of the brain and the absence of pigments allow to reach the brain more easily. Additionally the flexibility of SLM will allow to drive single or multiple cells in 3D.

The system uses the same configuration of 488 nm laser and SLM presented in Chap. 3 and Fig. 3.3, the upgrade being a different microscope objective (with a higher NA and longer working distance). In addition to the SLM illumination and imaging system we added two SPIM planes orthogonal to each other and to the microscope objective. The configuration and combination of the optical system built are presented in Fig. 4.5.

The advantage of having two SPIM planes at 90° angle is to avoid the limits of a single direction SPIM in zebrafish. A SPIM plane coming from the front of the fish nicely illuminates the forebrain and part of the mid-brain but scatters a lot after few hundreds of microns in depth. A SPIM plane coming from the side nicely illuminates the first few hundreds of microns all along the zebrafish side but the eyes hide an important region of the brain (the thalamus), which can only be reached from the front.

Once the sample is set in 2% agarose (more details in Sect. 3.2.1) and positioned at the correct height at the focal plane of the imaging microscope objective, the

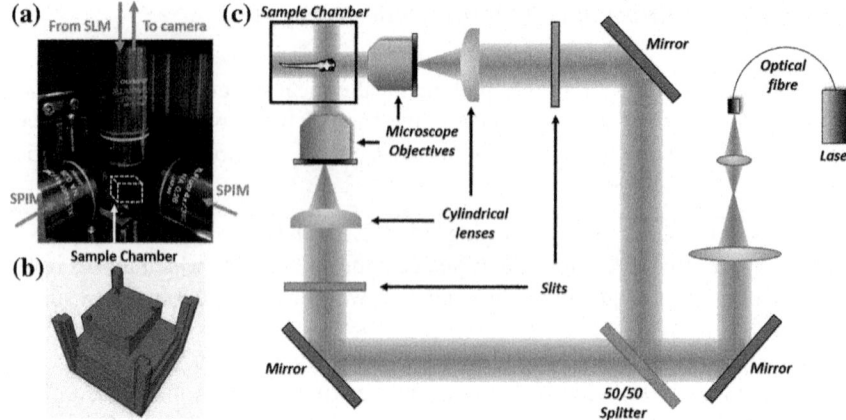

Fig. 4.5 Sketch of the set-up of the two SPIMs incorporated into the previous set-up. **a** Image of the overall configuration. **b** Chamber model designed with Tinkercad and latter printed on a 3D printer. 20 × 20 mm coverslips were glued on each side of the chamber for minimal SPIM distortions. **c** Details of the two SPIMS set-ups. For illumination, we used a 150 mW, 488 nm OBIS laser coupled to an optical fibre. The laser light was expanded and collimated A 50/50 splitter splits the beam into two beam with equal intensity. Each beam passes through a horizontal slit, a cylindrical lens and a 4×, 0.28 NA Olympus microscope objective. The two light sheets emerging from the microscope objectives focus down into the sample chamber where the zebrafish is set

brain activity can be imaged on the camera by illuminating the focal plane from the sides with the two SPIMs. The SLM is driven from the computer. The Labview program used to drive the SLM screen in Chap. 3 (Sect. 3.2.2) was calculating the hologram to display on the SLM screen to create multiple spots in 3D. A complication arises from targeting a region larger than a diffraction limited spot. A solution could be to scan rapidly the region of interest with a single diffraction limited spot. An interesting advantage would be a relatively low spreading of light along Z axis (beam propagation axis) compared to a large region illumination. In Fig. 4.6 I show that a single spot will illuminate a defined region in space with a FWHM defined by the NA of the microscope objective, however multiple simultaneous spots illumination create higher intensity in Z direction in the spots region causing the illumination to be much larger that the FWHM defined by the microscope objective predicts. By extension, a continuous large area of illumination has a large intensity spreading in Z.

As our fluorescent marker GCamp6f for neuronal activity is relatively fast (time-to-peak of about 50 ms) we would need to have a total scanning time of the region of interest below 50 ms. Our SLM has an image frame rate of 60 Hz, so a maximum of 60 single spots per second which is far too slow for the large volumes we want to illuminate.

For these reasons I decided to make a whole new program to drive the SLM, still using Labview but based on Gerchberg-Saxton algorithm.

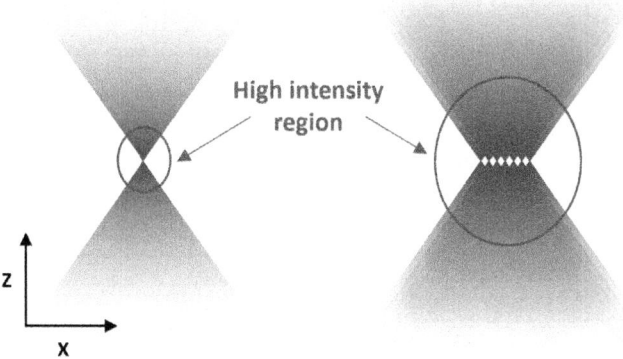

Fig. 4.6 Illustration of intensity for a single and multiple spots illumination. On the right multiple single spots intensity is the simple addition of the intensity from each individual spot which causes a higher spreading of intensity along the beam propagation

4.4.2 Set-Up Improvements Using SLM Flexibility

4.4.2.1 Introduction to Gerchberg-Saxton Algorithm

The Gerchberg Saxton (GS) algorithm has been found by R. W. Gerchberg and W. O. Saxton and was published in 1972 [47]. This method represents an iterative algorithm based on Fourier Transforms. From an initial intensity profile (laser profile) and a phase profile (SLM hologram) it calculates the theoretical desired intensity distribution, the next step is to compare it with the target intensity and to modify the phase profile until the target intensity is obtained. Initially, the phase profile image is made of random numbers and should converge after a certain number of iterations to the wanted phase profile to create the target intensity.

This algorithm has the advantage of creating any 2D illumination pattern, but has the disadvantage of needing time to calculate one hologram. Typically I found that 10–15 iterations are enough to create a very good approximation of the target image, and would require about 30 s to be calculated.

4.4.2.2 Implementing GS Algorithm on Labview

I made a user friendly Labview program which calculates the holograms to display on the SLM. In Fig. 4.7, I present the different options and flexibility that this program has. Because all the devices (laser, stage, camera etc.) are running with the software Micromanager and the SLM constructor doesn't provide any Micromanager driver, I chose to make a Labview program which can communicate with Micromanager via a text file. From Micromanager, one can write when to start the SLM display in the text file, and from the Labview program would permanently read this file. I

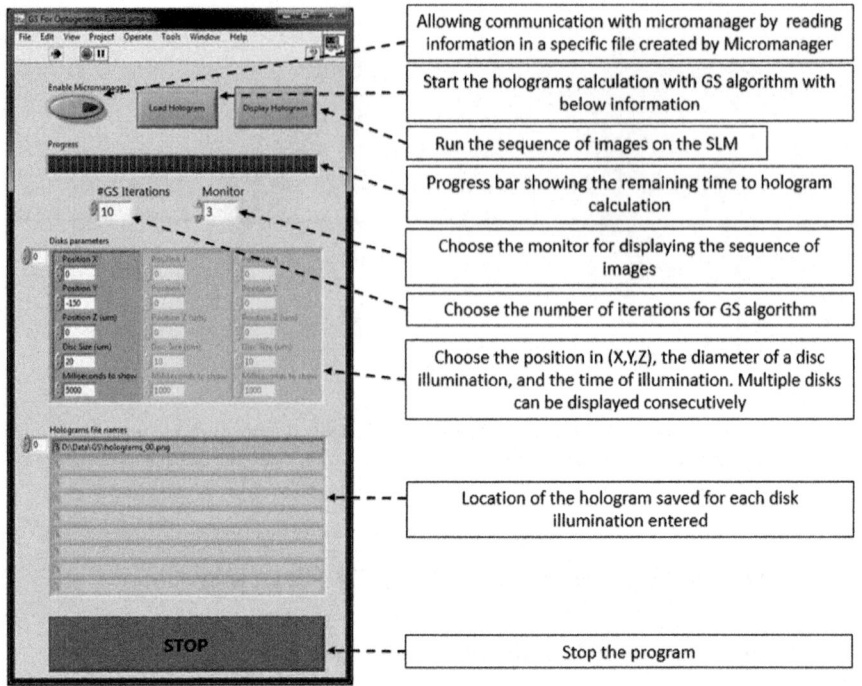

Fig. 4.7 Screen shot of the Labview program for hologram calculation using Gerchberg-Saxton algorithm. An "Enable Micromanager" button allows the permanent read of a .txt file edited by Micromanager. A "Load Hologram" button starts the calculation of the holograms with the parameters entered in the "Disk parameters" window. This calculation is based on Gerchberg-Saxton algorithm with a number of iterations entered in the "# GS Iterations" box. Finally the Monitor number can be chosen in "Monitor" box and the "Display Hologram" button start the display of the sequence of images calculated on the display chosen. This button can be triggered via Micromanager when the "Enable Micromanager" button is activated

calculated the delay time between Mincromanager editing the text file and the SLM display starting and it was less than 1 ms which is fast enough for our application.

In Fig. 4.7, we can see that this program creates disk images of various sizes and depths. Once all the parameters of the disk illumination are entered, it calculates the corresponding holograms using GS algorithm and save the sequence of images in a defined directory.

Since we are looking at brain regions and not single neurons, for simplicity we decided to create disk illuminations. After a set number of iterations the theoretical disk illumination is not perfectly smooth. Multiple disk sizes at different location were tested and an example of its performances is presented in Fig. 4.8.

In Fig. 4.8 we can see that the theoretical and the measured illuminations are not perfectly smooth. These irregularities should not however affect deep brain illumination as brain tissue will scatter and distort the incoming beam.

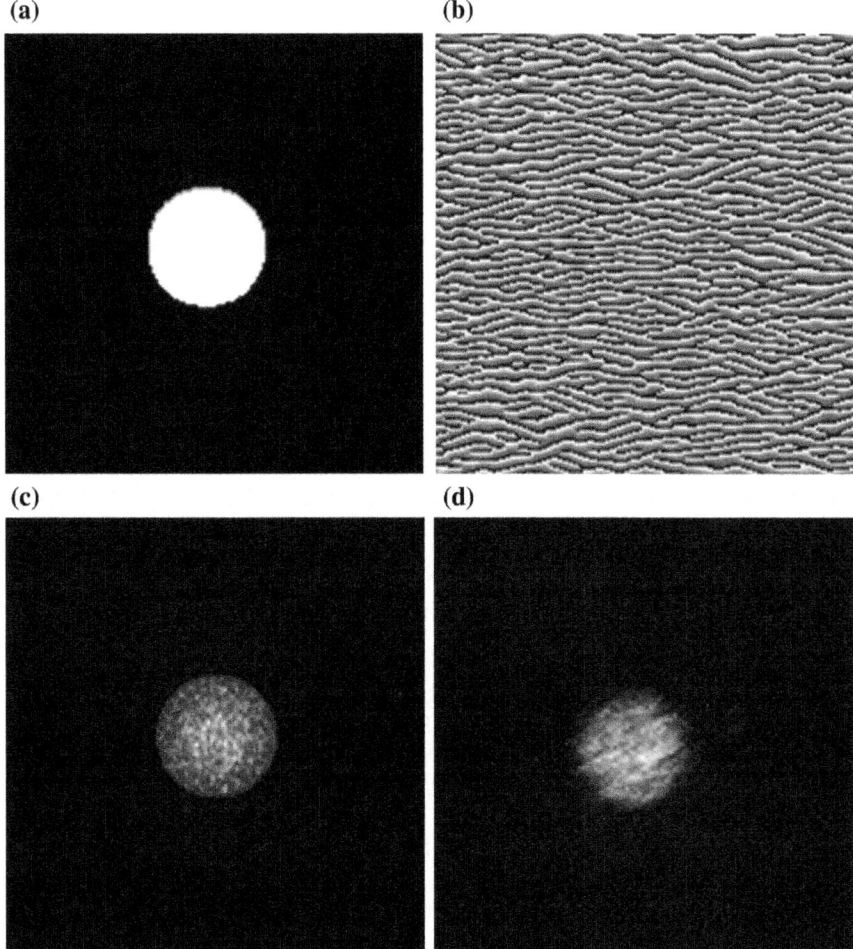

Fig. 4.8 Disk illumination with SLM. **a** Ideal input image: 30 μm disk. **b** Calculated hologram after 10 iterations. **c** Theoretical output image. **d** Measurement of the illumination from the camera after reflection on a microscope slide

4.4.3   Stimulation of Specific Brain Regions and Deduction of Brain Circuitry

With the optogenetics system described above, a wide range of different questions concerning the brain circuitry can be studied in zebrafish [48, 49]. To illustrate the current use of the Optogenetics system in our group, I present in this section a part of a study performed in zebrafish in vivo by Lucy Heap, in which I operated and maintained the holographic illumination equipment.

4.4.3.1 Summary of the Problem

The tectum in zebrafish is generally believed to be a visual structure, and models of its function have almost exclusively addressed inputs from the retina [50–52]. However, the mammalian superior colliculus, which is homologous to the tectum, receives inputs from various brain regions and sensory modalities [53]. The aim of the study is to explore nonretinal inputs to the tectum.

More precisely, the aim is to find out the excitatory and inhibitory tectal responses from various brain regions using optogenetics. We have studied three such brain regions: the cerebellum, thalamus and hypothalamus. In the context of this thesis only the cerebellum results will be presented.

4.4.3.2 Methods

GCamp6f, 6dpf animals were mounted dorsal side up in 2% low melt in a custom built glass sided imaging chamber (Fig. 4.5b) and were allowed to acclimate for 30 min prior to imaging. Optogenetic experiments were performed on the set-up presented in Fig. 4.5.

The brain imaging was performed at 5 Hz, for 70 s at a depth of 50 μm under the skin of the animal, with imaging focussed on the tectum. The activation of the cerebellum was performed at time points 50, 150 and 250, and lasted for 100 ms. The disk illumination was performed using holograms calculated with Gerchberg-Saxton algorithm (Sect. 4.4.2) and was focussed in the same z-plane as the imaging plane, 50 μm below the skin. In practice the SLM light efficiency and the scattering media would limit the depth reached by the SLM to trigger activity. In relatively low-scattering medium we can expect a 10 μm disc in the imaging to be very thin [41] however this thickness dramatically increases with the scattering properties of the medium and one photon lasers. By scanning a 10 μm disc in depth we determined that the light spreading in Z is about 15 μm above and below the disc imaging plane.

4.4.3.3 Results

The results presented in Fig. 4.9 show the location of the disk illumination (blue disk) in the cerebellar region and the corresponding excited cells in the tectum (in green) in response to the illumination. The neuronal responses over time have been represented in Fig. 4.9b and c and show the strong activity of excited neurons just after illumination. No inhibition f activity is observed.

These results prove that the larval zebrafish tectum receives inputs from the cerebellum. The same study has been made in the thalamus and hypothalamus in a very similar way (not show in the context on this thesis) and also show inputs from those regions. This suggest that the tectum has indeed a role in integration across multiple sensory modalities and is not exclusively a visual structure as previously thought.

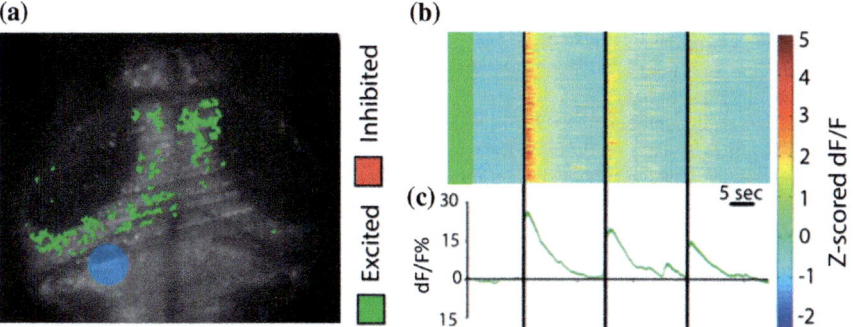

Fig. 4.9 Probing excitatory and inhibitory tectal responses to afferent stimulation. Analysis made and presented by Lucy Heap. **a** Cerebellar activation in a population of cerebellar neurons (shaded blue) resulted exclusively in excitatory responses in the tectum (shaded green). **b** Excited neuron activity over time. **c** Average response of all neurons in (b) over time (Color figure online)

4.5 Conclusions

In this chapter, I presented an overview of the different methods that have been developed over the past decades to study brain connectivity and function. I also showed that a wide range of combinations of optical systems exist to do optogenetics, and building our own system gave us the opportunity to conduct the specific study and answer specific, relevant questions.

Decoding the brain remains a very complex problem but can be broken down in a defined number of problems which can be investigated and solved individually. An example presented here addresses the relationship between the cerebellum and the tectum. With optogenetics, I determined that activation of the cerebellum leads, directly or indirectly, to excitation in the tectum.

In the next Chapter, I aim to gain further understanding of light's interactions with biological tissue, and look to address specific questions related to the perception and processing of vestibular stimuli.

References

1. D. Regan, *Human Brain Electrophysiology: Evoked Potentials and Evoked Magnetic Fields in Science and Medicine* (Elsevier, 1989)
2. A. Puce, D. Perrett, Electrophysiology and brain imaging of biological motion. Philos. Trans. R. Soc. Lond. Ser. B Biol. Sci. **358**(1431), 435 (2003)
3. D. Noble, The surprising heart: a review of recent progress in cardiac electrophysiology. J. Physiol. **353**, 1 (1984)
4. G. Burnstock, M.E. Holman, C.L. Prosser, Electrophysiology of smooth muscle. Physiol. Rev. **43**(3), 482 (1963)
5. M.W. Berns, Partial cell irradiation with a tunable organic dye laser. Nature **240**(5382), 483 (1972)

6. J. Sulston, J. White, Regulation and cell autonomy during postembryonic development of caenorhabditis elegans. Dev. Biol. **78**(2), 577 (1980)
7. O. Yizhar, L.E. Fenno, T.J. Davidson, M. Mogri, K. Deisseroth, Optogenetics in neural systems. Neuron **71**(1), 9 (2011)
8. K. Deisseroth, Optogenetics. Nat. Methods **8**(1), 26 (2011)
9. G. Nagel, T. Szellas, W. Huhn, S. Kateriya, N. Adeishvili, P. Berthold, D. Ollig, P. Hegemann, E. Bamberg, Channelrhodopsin-2, a directly light-gated cation-selective membrane channel. Proc. Natl. Acad. Sci. **100**(24), 13940 (2003)
10. F. Zhang, L.-P. Wang, E.S. Boyden, K. Deisseroth, Channelrhodopsin-2 and optical control of excitable cells. Nat. Methods **3**(10), 785 (2006)
11. A.B. Arrenberg, F. Del Bene, H. Baier, Optical control of zebrafish behavior with halorhodopsin. Proc. Natl. Acad. Sci. **106**(42), 17968 (2009)
12. B. Schobert, J.K. Lanyi, Halorhodopsin is a light-driven chloride pump. J. Biol. Chem. **257**(17), 10306 (1982)
13. S.A. Hires, L. Tian, L.L. Looger, Reporting neural activity with genetically encoded calcium indicators. Brain Cell Biol. **36**(1), 69 (2008)
14. Y. Zhao, S. Araki, J. Wu, T. Teramoto, Y.-F. Chang, M. Nakano, A.S. Abdelfattah, M. Fujiwara, T. Ishihara, T. Nagai, R.E. Campbell, An expanded palette of genetically encoded ca2+ indicators. Science **333**(6051), 1888 (2011)
15. T.-W. Chen, T.J. Wardill, Y. Sun, S.R. Pulver, S.L. Renninger, A. Baohan, E.R. Schreiter, R.A. Kerr, M.B. Orger, V. Jayaraman, L.L. Looger, K. Svoboda, D.S. Kim, Ultra-sensitive fluorescent proteins for imaging neuronal activity. Nature **499**(7458), 295 (2013)
16. H. Lutcke, M. Murayama, T. Hahn, D. Margolis, S. Astori, S. Meyer, W. Gobel, Y. Yang, W. Tang, S. Kugler, R. Sprengel, T. Nagai, A. Miyawaki, M. Larkum, F. Helmchen, M. Hasan, Optical recording of neuronal activity with a genetically-encoded calcium indicator in anesthetized and freely moving mice. Front. Neural Circuits **4**(9) (2010)
17. L. Tian, S.A. Hires, T. Mao, D. Huber, M.E. Chiappe, S.H. Chalasani, L. Petreanu, J. Akerboom, S.A. McKinney, E.R. Schreiter, C.I. Bargmann, V. Jayaraman, K. Svoboda, L.L. Looger, Imaging neural activity in worms, flies and mice with improved gcamp calcium indicators. Nat. Methods **6**(12), 875 (2009)
18. W. Gobel, F. Helmchen, In vivo calcium imaging of neural network function. Physiology **22**(6), 358 (2007)
19. T. Bozza, J.P. McGann, P. Mombaerts, M. Wachowiak, In vivo imaging of neuronal activity by targeted expression of a genetically encoded probe in the mouse. Neuron **42**(1), 9 (2004)
20. H.I.C. Dalgarno, T. Cizmar, T. Vettenburg, J. Nylk, F.J. Gunn-Moore, K. Dholakia, Wavefront corrected light sheet microscopy in turbid media. Appl. Phys. Lett. **100**(19), 191108 (2012)
21. M. Nixon, O. Katz, E. Small, Y. Bromberg, A.A. Friesem, Y. Silberberg, N. Davidson, Real-time wavefront shaping through scattering media by all-optical feedback. Nat. Photonics **7**(11), 919 (2013)
22. R. Bowman, A. Wright, M. Padgett, An slm-based shack-hartmann wavefront sensor for aberration correction in optical tweezers. J. Optics **12**(12), 124004 (2010)
23. T. Cizmar, M. Mazilu, K. Dholakia, In situ wavefront correction and its application to micromanipulation. Nat. Photonics **4**(6), 388 (2010)
24. B.A. Flusberg, E.D. Cocker, W. Piyawattanametha, J.C. Jung, E.L.M. Cheung, M.J. Schnitzer, Fiber-optic fluorescence imaging. Nat. methods **2**(12), 941 (2005)
25. D. Miyamoto, M. Murayama, The fiber-optic imaging and manipulation of neural activity during animal behavior. Neurosci. Res. **103**, 1 (2016)
26. S.C. Johnson, Optogenetics: an illuminating journey into the brain. Opt. Photon. News **22**(7), 26 (2011)
27. M. Ploschner, T. Tyc, T. Cizmar, Seeing through chaos in multimode fibres. Nat. Photonics **9**(8), 529 (2015)
28. T. Cizmar, K. Dholakia, Exploiting multimode waveguides for pure fibre-based imaging. Nat. Commun. **3**, 1027 (2012)

29. J. Huisken, J. Swoger, F. Del Bene, J. Wittbrodt, E.H.K. Stelzer, Optical sectioning deep inside live embryos by selective plane illumination microscopy. Science **305**(5686), 1007 (2004)
30. J. Huisken, D.Y.R. Stainier, Selective plane illumination microscopy techniques in developmental biology. Dev. (Cambridge, England) **136**(12), 1963 (2009)
31. R. Tomer, M. Lovett-Barron, I. Kauvar, A. Andalman, V.M. Burns, S. Sankaran, L. Grosenick, M. Broxton, S. Yang, K. Deisseroth, Sped light sheet microscopy: fast mapping of biological system structure and function. Cell **163**(7), 1796 (2015)
32. K. Greger, J. Swoger, E.H. Stelzer, Basic building units and properties of a fluorescence single plane illumination microscope. Rev. Sci. Instrum. **78**(2), 023705 (2007)
33. A.H. Voie, D.H. Burns, F.A. Spelman, Orthogonal-plane fluorescence optical sectioning: three-dimensional imaging of macroscopic biological specimens. J. Microsc. **170**(Pt 3), 229 (1993)
34. J. Buytaert, E. Descamps, D. Adrianens, J. Dirckx, *Orthogonal-Plane Fluorescence Optical Sectioning: A Technique for 3-D Imaging of Biomedical Specimens* (Formatex Research Center, 2010)
35. T. Vettenburg, H.I.C. Dalgarno, J. Nylk, C. Coll-Llado, D.E.K. Ferrier, T. Cizmar, F.J. Gunn-Moore, K. Dholakia, Light-sheet microscopy using an airy beam. Nat. Methods **11**(5), 541 (2014)
36. E.H.K. Stelzer, Light-sheet fluorescence microscopy for quantitative biology. Nat. Methods **12**(1), 23 (2015)
37. F.O. Fahrbach, F.F. Voigt, B. Schmid, F. Helmchen, J. Huisken, Rapid 3d light-sheet microscopy with a tunable lens. Optics Express **21**(18), 21010 (2013)
38. M.B. Bouchard, V. Voleti, C.S. Mendes, C. Lacefield, W.B. Grueber, R.S. Mann, R.M. Bruno, E.M.C. Hillman, Swept confocally-aligned planar excitation (scape) microscopy for high-speed volumetric imaging of behaving organisms. Nat. Photonics **9**(2), 113 (2015)
39. M. Weber, J. Huisken, Light sheet microscopy for real-time developmental biology. Curr. Opin. Genet. Dev. **21**(5), 566 (2011)
40. S. Quirin, J. Jackson, D.S. Peterka, R. Yuste, Simultaneous imaging of neural activity in three dimensions. Front. Neural Circuits **8**, 29 (2014)
41. C. Lutz, T.S. Otis, V. DeSars, S. Charpak, D.A. DiGregorio, V. Emiliani, Holographic photolysis of caged neurotransmitters. Nat. Methods **5**(9), 821 (2008)
42. Y. Takiguchi, T. Otsu, T. Inoue, H. Toyoda, Self-distortion compensation of spatial light modulator under temperature-varying conditions. Optics Express **22**(13), 16087 (2014)
43. R. Heintzmann, Correcting distorted optics: back to the basics. Nat. Methods **7**(2), 108 (2010)
44. E. Papagiakoumou, V. de Sars, D. Oron, V. Emiliani, Patterned two-photon illumination by spatiotemporal shaping of ultrashort pulses. Optics Express **16**(26), 22039 (2008)
45. A. Badura, X.R. Sun, A. Giovannucci, L.A. Lynch, S.S.H. Wang, Fast calcium sensor proteins for monitoring neural activity. Neurophotonics **1**(2), 025008 (2014)
46. J. Mertz, Optical sectioning microscopy with planar or structured illumination. Nat. Methods **8**(10), 811 (2011)
47. R.W. Gerchberg, W.O. Saxton, A practical algorithm for the determination of the phase from image and diffraction plane pictures. Optik (Jena) **35**, 237–246 (1972)
48. F. Del Bene, C. Wyart, Optogenetics: a new enlightenment age for zebrafish neurobiology. Dev. Neurobiol. **72**(3), 404 (2012)
49. R. Portugues, K.E. Severi, C. Wyart, M.B. Ahrens, Optogenetics in a transparent animal: circuit function in the larval zebrafish. Curr. Opinion Neurobiol. **23**(1), 119 (2013)
50. L.M. Nevin, E. Robles, H. Baier, E.K. Scott, Focusing on optic tectum circuitry through the lens of genetics. BMC Biol. **8**(1), 126 (2010)
51. P. Sajovic, C. Levinthal, Visual cells of zebrafish optic tectum: mapping with small spots. Neuroscience **7**(10), 2407 (1982)
52. M. Leslie, How the optic tectum stacks up. J. Cell Biol. **211**(4), 719 (2015)
53. M.A. Meredith, B.E. Stein, Visual, auditory, and somatosensory convergence on cells in superior colliculus results in multisensory integration. J. Neurophysiol. **56**(3), 640 (1986)

Chapter 5
Investigation of Optical Properties of Otoliths with Optical Trapping

Investigating the functioning of biological system often ask to, not only image the different processes and elements involved, but also manipulate them to find information that could not be found visually, such as forces of motion, strength of bondings, elasticity, viscosity to cite a few.

In the previous chapters, I have presented how optical tools can contribute to the mapping of neural function. The general approach is to present a sensory stimulus or perform an optogenetic manipulation, and to observe the responses of neurons throughout the brain. For some questions, however, stimuli are incompatible with calcium imaging. This is the case for the vestibular system, in which stimulation requires the physical movement of the animal. To get around this limitation, we are considered applying optical trapping forces to the ear stones of larval zebrafish, in order to produce a fictive vestibular stimulus in a stationary animal. In this chapter, I perform a thorough characterisation of the ear stone's optical properties to see whether such in-vivo trapping is plausible.

These results are currently under review here:

I. A. Favre-Bulle, A. B. Stilgoe, H. Rubinsztein-Dunlop, and E. K. Scott. *Optical trapping of otoliths drives vestibular behaviours in larval zebrafish*. Under review in Nature Communications (2017).

5.1 Description of the Vestibular System

The vestibular sensory organs in vertebrates are complex systems. They are constituted of semicircular canals, which are sensitive to rotation, and otoliths, or ear stones, which are responsible for balance and hearing. Semi-circular canals are three canals interconnected and oriented along the pitch, roll, and yaw axes [2] and therefore are able to detect flow direction and provide sensory input for any rotary movements.

© Springer International Publishing AG, part of Springer Nature 2018
I. A. Favre-Bulle, *Imaging, Manipulation and Optogenetics in Zebrafish*,
Springer Theses, https://doi.org/10.1007/978-3-319-96250-4_5

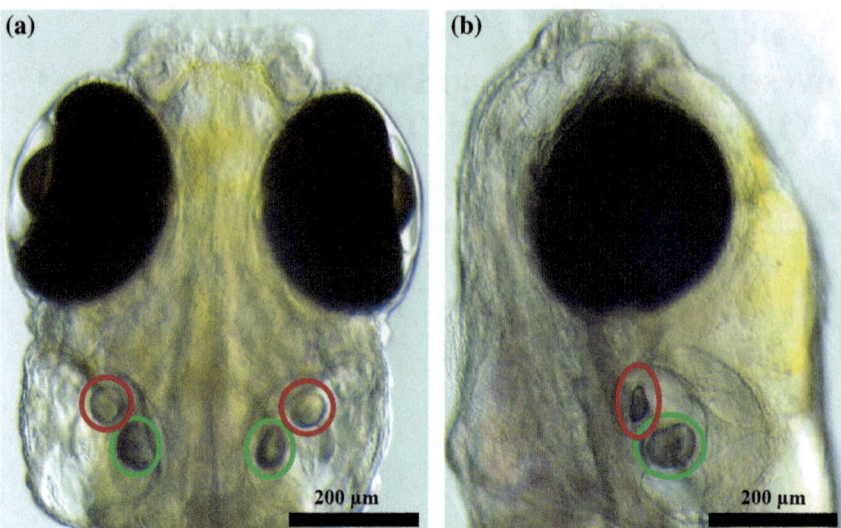

Fig. 5.1 Otolith location in a 6 dpf larvae zebrafish. Red circles show urticular otoliths and green circle saccular otoliths. Top view in (**a**) and side view in (**b**)

Otoliths are calcium carbonate crystals. There are a total of two, one in each ear, in young larval zebrafish: the anterior, or urticular otolith, is believed to be responsible for balance sensing only and the posterior, or saccular otolith, for hearing [3–5]. Otoliths are attached to the ear membrane with hair cells or neuronal receptor cells and therefore represent the starting point of the vestibular information processing.

Utricular otoliths are located about 150–200 μm deep under the skin, and are about 50–60 μm in diameter for the urticular otolith, and 70–90 μm for the saccular otolith in 6 days post fertilized (dpf) larvae. Figure 5.1 presents the top view and side view pictures of a 6 dpf larvae imaged on a bright field microscope. A total of 5 larvae have been imaged to get a reasonable approximation of otolith dimensions.

5.1.1 Necessity of a Non-invasive System

Because semicircular canals are not yet functional in larval zebrafish [6], they provide a simplified version of most vertebrate vestibular information processing. This implies that only the utricular otoliths can detect acceleration stimuli, as the saccular otolith is involved in hearing [4, 5], which makes the possibility of non-invasive manipulation of the utricular otolith very attractive for the investigation of the acceleration encoding.

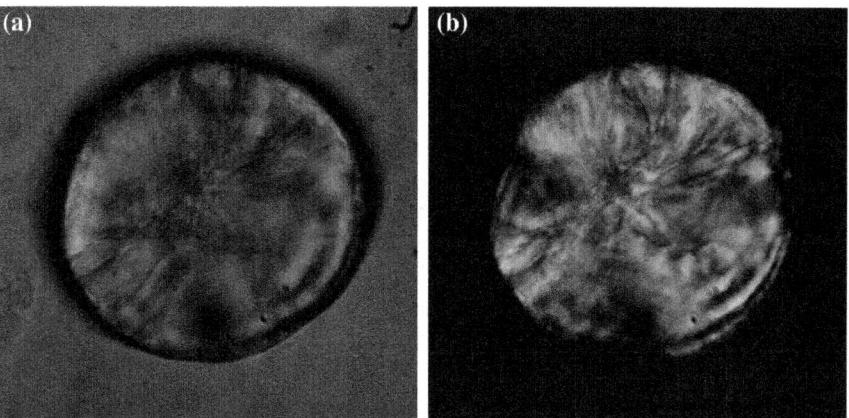

Fig. 5.2 Birefringence of Otolith. **a** Otolith imaged throught a single poraliser. **b** Otolith imaged throught crossed polarisers

Isolating the acceleration to the anterior otolith is, however, a very difficult task. Previous studies on acceleration sensing involve the development of a complex translational robot [7] or rotating platform [8, 9], but such techniques are disruptive as the whole fish body is accelerated. This also implies that the behavioural response can be difficult to extract, as the fish body encounters changing forces due to changing water flow or gravity, which will affect the behavioural response too.

Optical tweezers can apply very localised forces, and could isolate the otolith motion from the rest of the body, to drive a behavioral response. However, to investigate the possibility of non-invasive manipulation techniques on otolith, it is essential to know their optical properties.

5.1.2 Structure of the Anterior Otolith

Otoliths are composed primarily of crystalline calcium carbonate [10] and due to their aragonite structure they are birefringent and transparent. The refractive indices of aragonite crystal are $n_\alpha = 1.53$, $n_\beta = 1.68$, $n_\gamma = 1.69$ [11].

I observed this birefringence on otoliths dissected and set free in water. The otolith was placed in between two crossed polarisers and was illuminated on a bright field microscope. Figure 5.2 shows the appearance of a dark cross which is indicative of birefringence. This dark cross is not perfectly clear which suggests imperfections in the structure of the crystal, especially around the centre.

Despite their large size and mass, otoliths' transparency and high refractive index make them very good candidates for optical manipulation.

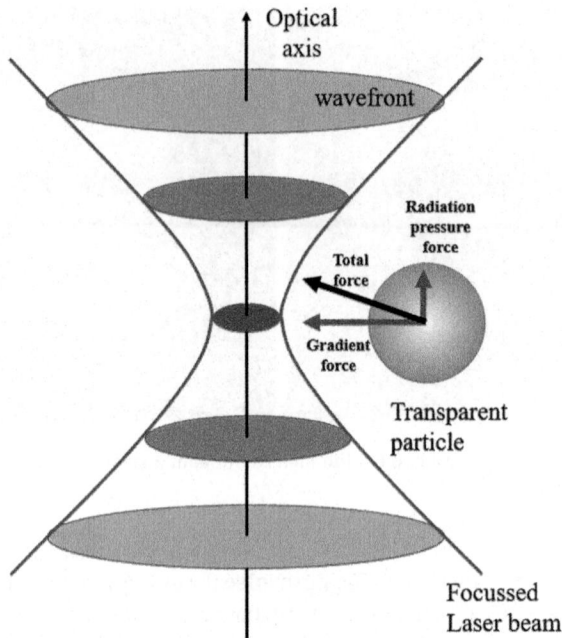

Fig. 5.3 Sketch of forces applied on a particle near a focussed laser beam. The gradient force attract a particle towards the region with the highest intensity. The radiation pressure force repel the particle in the direction of the beam propagation. The sum of those two force gives the total force towards a single stable position on the optical axis

5.2   Optical Trapping for Otolith Manipulation

5.2.1   Introduction to Optical Trapping

Ashkin's demonstration of optical trapping [12] is based on momentum transfer of light. A particle with a different refractive index to that of surrounding media being exposed to light will re-emit the light at a different angle. The result is momentum transfer, which places a physical force on the refracting particle. When the force is strong enough, and the particle light enough, this causes the particle to move.

Figure 5.3 sketches out the forces applied on a transparent particle in the vicinity of a focussed laser beam. In this case, Ashkin explained the phenomenon using two forces: the scattering force in the direction of the incident power and the gradient force in the direction of the intensity gradient. The gradient force is the result of the inhomogeneity of the laser intensity in space, it attracts the particle in the direction of higher intensity regions which is the focal point for a the focussed beam. The scattering force as its name indicates is the result of the multiple scattering events happening on the particle surface and resulting in a force counteracting the gradient force in the direction of beam propagation. In general, the particle properties allow the gradient force to be greater than the scattering force which results in a stable trap slightly offset from the center of the focussed beam in the beam propagation direction. In some cases, the scattering forces are such that no stable trap position exists.

From here on, I will refer to the optical force as the gradient force. I will not consider the scattering force, as I will be analysing movements and forces in the plane perpendicular to the beam propagation only.

An example of the gradient force for a spherical particle away from the centre of a focussed beam is presented in Fig. 5.3. When the optical trap is located very close to the centre of the particle, the force attracting it towards the centre is approximately linear. Outside this region, stronger forces appear which correspond to refraction at the edges of the particle. Taking into account this observation, we can deduce that the regions in the otolith where the strongest force could be applied are also its edges. This needs to be refined and the proper measurement of forces needs to be made in order to determine if the forces applied to the otolith can be sufficient to simulate movement.

5.2.2 Investigation of the Possibility of Otolith Manipulation

The first questions we want to answer is the possibility of optical traps to provide sufficient forces on otolith to simulate motion.

Otolith being typically around $r = 22.5\,\mu m$ in radius at the age we are interested in (6 dpf), they are therefore very large for common micro-manipulation systems, which typically work on spherical particles about 1–5 μm. Moreover due to its size, the otolith's weight is not negligible. Indeed, the acceleration experienced by the otolith while the fish is in movement is the key parameter which dictates the intensity of the response.

To get a rough figure for the acceleration perceived by the otolith when the fish is accelerating, we first observed few videos of free swimming zebrafish in water at 6 dpf. We calculated that their maximum speed during scooting was $24 \pm 5\,mm \cdot s^{-1}$ which was approximated to be an acceleration of $a = 240\,mm \cdot s^{-2}$. Moreover, as the calcium carbonate present in the otolith represents more than 95% of the weight of the otolith [10] by approximation we calculated its weight from the density of calcium carbonate in the aragonite structure which is $\rho = 2.83.10^{-12}\,g \cdot \mu m^{-3}$. The otolith mass M calculated can be expressed as the product of its volume and volume density ρ follows:

$$M = \frac{4}{3}\pi r^3 \rho. \tag{5.1}$$

And was calculated to be approximately $M = 64$ pg. From this we can deduce the corresponding force F on the otolith:

$$F = Ma. \tag{5.2}$$

(a)   (b)   (c)

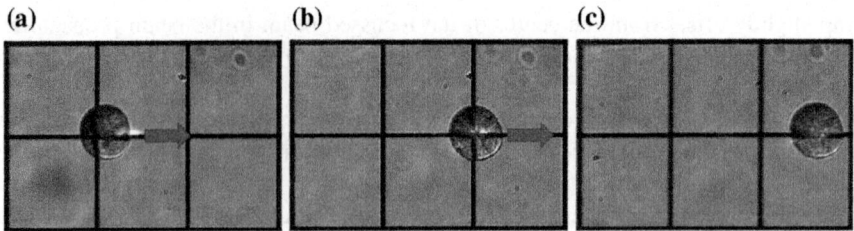

Fig. 5.4 a–c Snapshots of the moving otolith video with 100 mW of 1064 laser power focussed on the extreme right side of the otolith. Grid size is 100 μm

This rough calculation gives a total force of $F = 16$ pN. Optical trapping can indeed provide forces in the order of pN which strongly favour the possibility of optical manipulation.

5.2.3 Manipulation of Free Otolith

The next questions I investigated was whether free otoliths can be manipulated with an optical trap. To do so, I extracted multiple otoliths from zebrafish ears using sharp mechanical tweezers. The extraction was first made in E3 medium, the medium usually used for zebrafish constituted of salt water and methylene to prevent fungal growth. After extraction, they were left on a microscope slide with a drop of E3. The sample was then placed under a water immersed microscope objective. No dissolving or any modification could be observed after one hour.

The first trials revealed that the otoliths could be extracted entirely and freed from any residual tissue. The manipulation and displacement with optical traps was successful despite the tendency of the otolith to stick to the slide.

To reduce the stickiness of the otolith, I had to change the sample preparation procedure. The extraction of otolith was still done with sharp mechanical tweezers but the solution the otolith was left in was changed. After extraction, the otolith was soaked for one second in Tris-EDTA (10 mM Tris-HCl, 1 mM disodium EDTA), and just after placed in a chamber filled with 0.05% Trypsin-EDTA (gibco, Life Technologies). The chamber was made of one microscope slide at the bottom, two layers of double sided adhesive tape for the walls, and a thin coverslip on top. A 100 μm grid was printed on the top side of the microscope slide. This chamber needed to be closed due to the high perturbation in the solution and the constant loss of the otolith when the microscope objective was approached and immersed in the sample. This procedure resulted in much easier manipulation of otolith along X and Y but no lift along Z axis was possible. With this method a total laser power of 100 mW at the otolith was sufficient to manipulate the otolith compared to 300 mW for the previous procedure. Snapshots of the videos are present in Fig. 5.4.

Laser powers of 100–300 mW may be too high for a number of biological samples, however, in Chap. 6, I present consistent behavioural responses of the simulation of side movements with up to 600 mW of laser power focussed in the otolith with the same laser and microscope objective. The fish may perceive the heat associated with the laser, but as long as it is not sufficiently noxious to elicit a behavioural response, this remains a promising approach for delivering simulated acceleration.

5.3 Force Measurement with Optical Trapping

At this point, it became important to identify the positions within the otolith that produced the greatest trapping forces. Indeed, going in-vivo will introduce a lot of distortions due to scattering which would result in a distorted focussed beam and a weaker trapping force. This is why it is essential to precisely locate those regions to give higher chances of successful acceleration simulation experiments.

5.3.1 Introduction to Light Deflection Method

One way to measure optical forces experienced by a transparent particle is to measure the amount of light deflected by this particle [13]. In Fig. 5.5 I present an illustration of light deflection by a particle slightly displaced from the focal plane. Light deflection is proportional to momentum transfer to the particle. In the linear regime the force along one axis F_x is proportional to its position x:

$$F_x = -kx, \tag{5.3}$$

where k is a constant factor. The main idea here is that the amount of light deviated can be measured via an adequate light collecting device and these measurements provide the possibility to calculate the particle position and deduce the force. From the previous section, we know that the particle position in the trap dictates the force applied on the particle by the trap which means that by placing the optical trap at different position in a particle, or by scanning a particle with an optical trap, we can calculate the distribution of forces throughout the sample.

Theoretically, we can calculate the distribution of forces within an otolith by approximating it as a sphere and use ray optics to model the propagation of rays of light. Ray optics calculations can be applied when the wavelength of the incident beam is small compared to the object's size, as long as the focus of any beam constructed out of rays is far from the interface between the object and the medium that surrounds it. For this situation, the ray optics model defines the incoming beam as a Gaussian distribution of rays [14, 15] as follows:

$$U_{(r,z)} = \frac{1}{w(z)} \left(\frac{\sqrt{2}r}{w(z)} \right)^l e^{(-r/w(z))^2}, \tag{5.4}$$

where:

$$w(z) = w_0\sqrt{1 + (z/z_r)^2}, \tag{5.5}$$

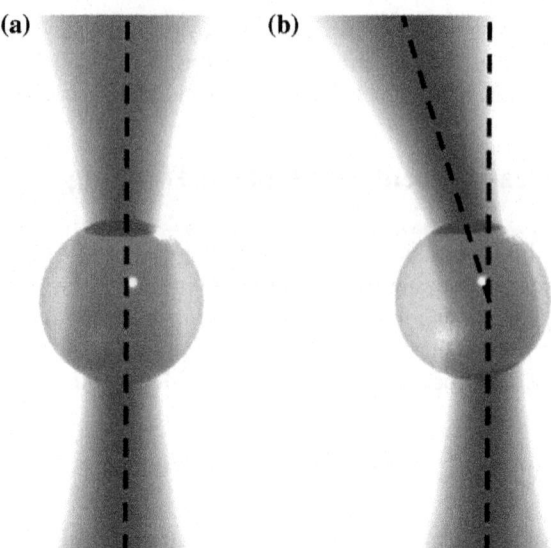

Fig. 5.5 Sketch of a focussed beam deflected by a transparent particle. **a** A particle in the centre of the beam does not deflect the beam away from the incoming beam direction. **b** A slight displacement of the particle orthogonally to the incoming beam deflects the light noticeably

and where z is the displacement along the beam propagation direction, w_0 is the beam waist at the focus, r is the transverse position and z_r is the Raleigh range (defined in Sect. 2.2).

To find the forces experienced by a transparent sphere, only the angle between the external ray and the surface normal of the sphere and the angle of the same ray inside the sphere need to be known to calculate the force contribution of each ray. Summing up all the rays' contribution would result in the calculation of the total force applied on the sphere. However, it is impossible experimentally to collect all the rays coming out from an object; a limited angle of collection is dictated by condenser lens or any other light collection optics.

Therefore it is important to consider in the calculation the light that is lost due to limited collection angle. Figure 5.6a presents an example of ray after multiple reflection in a transparent sphere, showing that multiple scattering leads to high scattering angles hardly collectable by a condenser lens. Figure 5.6b shows the translation of those ray vector to a centre point showing their wide distribution and the inevitable limited performances of condensing lenses. Finally, Fig. 5.6c shows the comparison of the total force calculated considering all the forward scattered rays being collected and the forces calculated when limited by collection angle. In the case calculated here, the difference between the ideal forward scattering and the limited collection area measurement of force is clear. The maximum peak force is about 10% different, which is not negligible and should be taken into account in the measurements. Also, the reduced collection area force has the opposite sign (dashed purple ellipses) and 3 points of zero force (green crosses) which do not represent three stable particle position but only are the result of the limited ray angles collection.

The force acting on a particle trapped in optical tweezers can be expressed as:

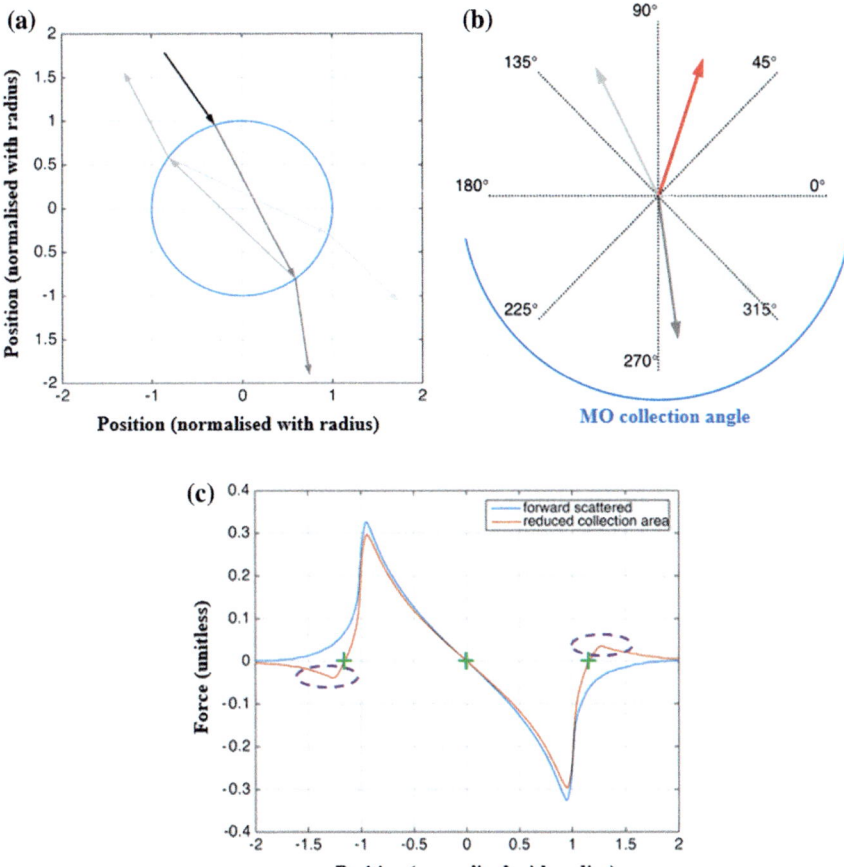

Fig. 5.6 Force calculation based on ray optics method. **a** Example of a ray encountering multiple scattering events within a particle. Arrow shade represents the loss of ray intensity as the ray reflects and refracts inside or outside the sphere. **b** An example of measurement from a condenser lens effectively collects the light from the region drown in blue. The lens chosen has a numerical aperture of 1.33 in water. All the rays cannot be collected, hence it is an incomplete measurement. Arrows show that for this particular ray most of the forward scattered light falls within the collection region. **c** Difference between the ray optics model when all the forward scattered light is collected (blue) and only light within the field view of the microscope is captured (red). The reduced collection area force present opposite sign values (dashed purple ellipses) and 3 zero force points (green crosses) which are not the consequences of limited angles collection

$$F = \frac{nP}{c}Q \qquad (5.6)$$

where F is the force in Newtons, c is the speed of light in vacuum, n is the refractive index, P is the radiant flux of light, or optical power, and Q is the force contributed per photon momentum. This implies that the force is proportional to the quality factor Q and the laser power. Also its maximum, which is on the order of 5×10^{-10}

N for a 500 mW laser, occurs when the trap is just before the sphere's edges, and this force decreases dramatically when the trap moves from the edge of the otolith in either the positive or negative direction. These results suggest that the optical trap or focussed beam would have to be precisely positioned in the otolith to maximize the force applied to it. Indeed, from Fig. 5.6c we found that the force drops by 10% with a shift of less than 0.5 μm away from the optimal position.

These modelling results lead us to a better understanding of the limits of our system and the level of uncertainty in measurements performed on otoliths in an experiment.

5.3.2 Experimental Otolith Scanning and Force Measurements

The experimental measurements of optical scanning based on the method described above were performed using the microscope configuration presented in Fig. 5.6 and built by Alexander Stilgoe. Motorized control of the stage was achieved using a custom LabVIEW interface to coordinate the movement of the stage and data capture. A schematic of this approach is shown in Fig. 5.7.

The set-up is built such that an IR fibre laser (1064 nm) creates an optical trap in the sample chamber via a water immersed microscope objective. By moving the sample chamber with the motorized stage we scanned multiple otolith on a grid of 80 by 80 positions in the centre plane of the otolith.

After passing through the sample, the scattered light is detected by the position sensitive detector (PSD). The PSD signals are converted to transverse forces (F_X and F_Y) with a predetermined scaling factor to convert volts into Newtons. Finally the total force is calculated as the amplitude of the vector sum:

$$F = \sqrt{F_X^2 + F_Y^2}. \tag{5.7}$$

An example of forces measured in an utricular otolith scan is presented in Fig. 5.8. The edges of the otolith are easily visible in the X and Y direction. We can also see that the force profile with distance calculated and presented in Fig. 5.6c for reduced collection area are similar to those in Fig. 5.8d and e with random spikes possibly originating from the crystal imperfections. Surprisingly, the imperfections or radial cracks visible in the wide field illumination (Fig. 5.8f) are also visible in the optical scan (Fig. 5.8c) which suggests that they have a non negligible contribution to the total force experienced by the otolith.

In addition, we can see that the force measured around the otolith edges are not perfectly smooth all around the otolith, but some radial directions have stronger forces than others. In Fig. 5.9a, we can see a 3D plot of the total force which shows two high intensity region and two lower intensity regions. This is a consequence of the birefringence observed and described in Sect. 5.1.2. From Fig. 5.9b we have determined that the two wide regions have stronger forces of 5.0 ± 1.5 pN, and two narrower regions 3.0 ± 1.5 pN.

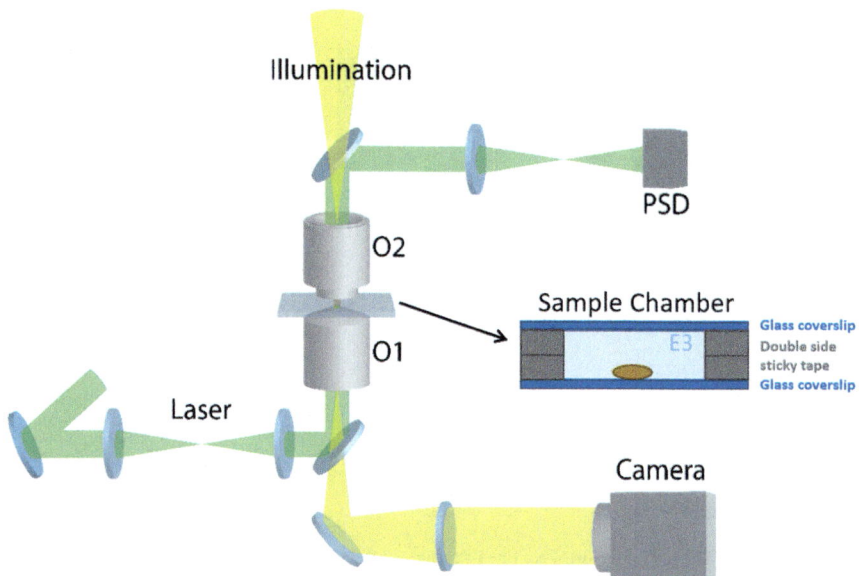

Fig. 5.7 Sketch of the set-up used for otolith scanning. A fibre laser (1070 nm IPG Photonics YLD-5) was focused by a water immersion objective lens O1 (Olympus UPLSAPO60Xw) to a limited diffracted spot, and a scanning nano stage (PI P-563.3CD) was used to move the otolith. The light deflected by the otolith was collected by a second silicone oil immersion objective lens O2 (Olympus UPLSAPO100XS) and imaged by a transfer lens onto a position sensitive detector PSD (On-track PSM2-10 with OT-301DL amplifier). The sample chamber is made of two glass coverslips ensuring flat surfaces along the beam propagation direction. Walls were made out of double side sticky tape to avoid medium E3 leaks

Another interesting point is the radial behaviour of the force. In Fig. 5.9b, red arrows represent the direction and strength of the force, we can clearly see that on the edges the total force is directed towards the centre. To emphasize this point in Fig. 5.9c we plot the force X (in red) and Y axis (in green) for $X = 0$. The red curve varies clearly very little at the otolith edges region which explains a radial force. However, it is not negligible in the centre region and the imperfections of the crystal create non zero forces which would appear as random from otolith to otolith.

These results show that the birefringence has a significant effect on the total force exerted on the otoliths, however, this effect is moderate compared to the effect of structural heterogeneity of crystal imperfections within the otolith. From analysis of the sharp and narrow regions of strong forces, we can conclude that the trap force is much more dependent on the placement of the focal point near the edge of the otolith than on its structural properties. Moreover, OTs at the edge of the otolith consistently produced radial forces along the X and Y axes which fits well with the modelling even though it did not consider heterogeneity and birefringence.

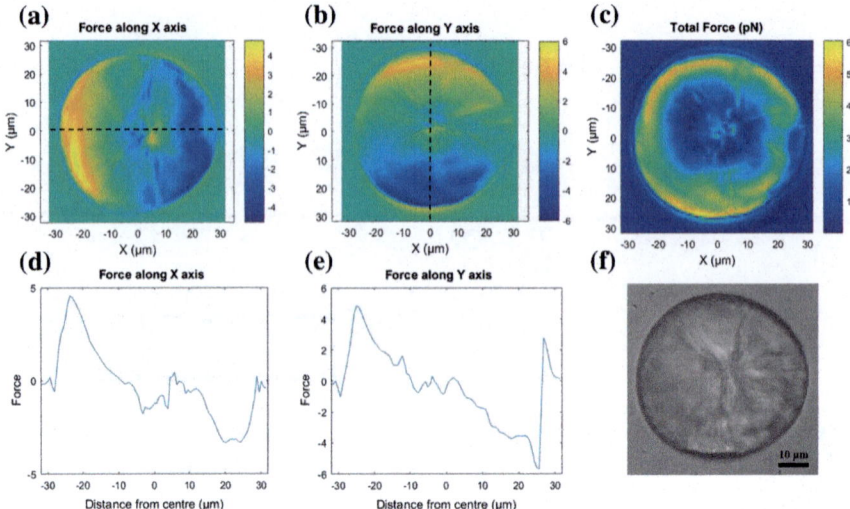

Fig. 5.8 Forces (in pN) measured by the PSD for a 80×80 grid along X axis in (**a**) and Y axis in (**b**). **c** Total force calculated from (**a**) and (**b**). **d** Force along X axis at $Y = 0$ (in pN). Also represented in black dashed line in (**a**). **e** Force along Y axis at $X = 0$ (in pN). Also represented in black dashed line in (**b**). **f** Wide field image of the otolith scanned showing internal imperfections

5.4 Discussion

Having investigated the optical properties of the utricular otoliths for OT, I showed that the calcium carbonate crystal which constitutes more than 95% of the otolith makes it birefringent. This birefringence would affect the force produced by a focussed polarised beam if it was a perfect crystal, however, the images of otolith obtained with wide field illumination and the measurements of OT scanning prove that the random polycrystalline structure affects the force as much as the birefringence to polarised light. This lead us to focus effort on the determination of the optimal maximum force location rather than the choice of polarization.

From the OT scanning, we showed that optical forces in the order of pN are achievable, which is what an otolith would encounter during an average fish acceleration. These results yield an upper limit of the force we would apply in-vivo, as the focussed beam will be distorted due to highly heterogeneous tissue between the coverslip and the otolith which will result in weaker forces.

The next step would be to test the OT technique in-vivo and see if we can indeed apply sufficient force in-vivo to simulate acceleration and trigger behavioural responses. Zebrafish are transparent, so absorption should be low, but it is possible

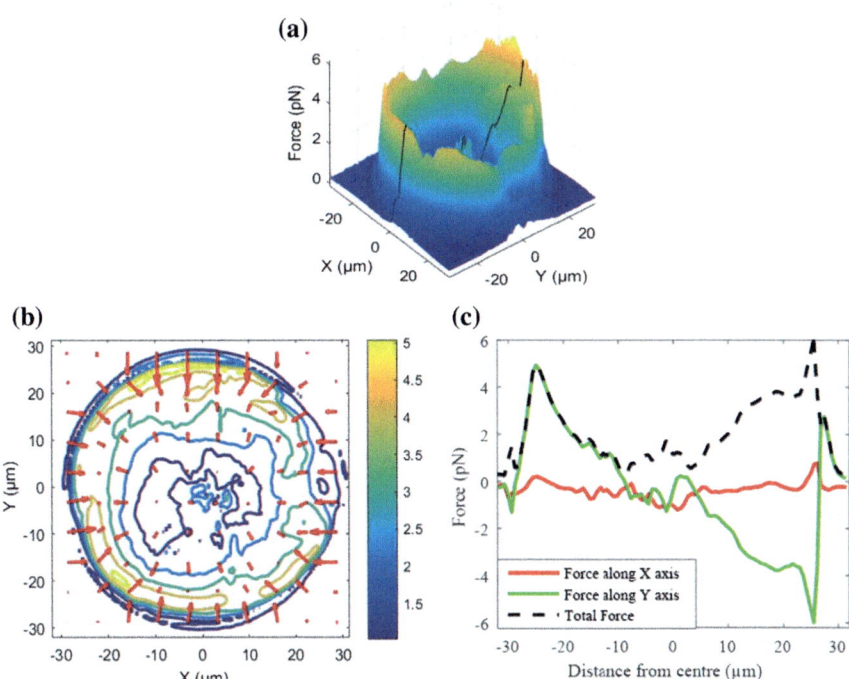

Fig. 5.9 Direction and strength of optical forces on otolith. **a** 3D plot of total force with position. **b** Force contours (scale yellow to blue in pN) and direction (red arrows) within the otolith. **c** Example of force at $X = 0$ along the X axis (red curve) and Y axis (green curve) showing the weak contribution of the Force along X at $X = 0$ to the total force(dashed black line) which explains the radial force

that refractive index heterogeneity in the animal will reduce the trap's power to the point where it would both be effective.

References

1. I.A. Favre-Bulle, A.B. Stilgoe, H. Rubinsztein-Dunlop, E.K. Scott, Optical trapping of otoliths drives vestibular behaviours in larval zebrafish. Under review in Nature Communications (2017)
2. M.M. Bever, D.M. Fekete, Atlas of the developing inner ear in zebrafish. Dev. Dyn. **223**(4), 536 (2002)
3. K.E. Cullen, The vestibular system: multimodal integration and encoding of self-motion for motor control. Trends Neurosci. **35**(3), 185 (2012)
4. M. Inoue, M. Tanimoto, Y. Oda, The role of ear stone size in hair cell acoustic sensory transduction. Sci. Rep. **3** (2013)
5. B.B. Riley, S.J. Moorman, Development of utricular otoliths, but not saccular otoliths, is necessary for vestibular function and survival in zebrafish. J. Neurobiol. **43**(4), 329 (2000)
6. J.C. Beck, E. Gilland, D.W. Tank, R. Baker, Quantifying the ontogeny of optokinetic and vestibuloocular behaviors in zebrafish, medaka, and goldfish. J. Neurophysiol. **92**(6), 3546 (2004)

7. F. Bonnet, P. Retornaz, J. Halloy, A. Gribovskiy, F. Mondada, Development of a mobile robot to study the collective behavior of zebrafish, in *2012 4th IEEE RAS EMBS International Conference on Biomedical Robotics and Biomechatronics (BioRob)*, pp. 437–442 (2012)
8. I.H. Bianco, L.H. Ma, D. Schoppik, D.N. Robson, M.B. Orger, J.C. Beck, J.M. Li, A.F. Schier, F. Engert, R. Baker, The tangential nucleus controls a gravito-inertial vestibulo-ocular reflex. Curr. Biol. **22**(14), 1285 (2012)
9. W. Mo, F. Chen, A. Nechiporuk, T. Nicolson, Quantification of vestibular-induced eye movements in zebrafish larvae. BMC Neurosci. **11**(1), 1 (2010)
10. E.C. Steven, Chemistry and composition of fish otoliths: pathways, mechanisms and applications. Mar. Ecol. Prog. Ser. **188**, 263 (1999)
11. W.L. Bragg, The refractive indices of calcite and aragonite. Proc. R. Soc. Lond. A. **105**(732), 370 (1924). Series A, Containing Papers of a Mathematical and Physical Character
12. A. Ashkin, Acceleration and trapping of particles by radiation pressure. Phys. Rev. Lett. **24**, 156 (1970)
13. A. Farre, F. Marsa, M. Montes-Usategui, Optimized back-focal-plane interferometry directly measures forces of optically trapped particles. Opt. Express **20**(11), 12270 (2012)
14. L. Allen, M.W. Beijersbergen, R.J.C. Spreeuw, J.P. Woerdman, Orbital angular momentum of light and the transformation of Laguerre Gaussian laser modes. Phys. Rev. A **45**(11), 8185 (1992). PRA
15. M. Bohmer, J. Enderlein, Orientation imaging of single molecules by wide-field epifluorescence microscopy. J. Opt. Soc. Am. B **20**(3), 554 (2003)

Chapter 6
Optical Manipulation of Otoliths In-Vivo

Despite the light scattering that occurs in biological tissues, in-vivo optical trapping is possible, and has been demonstrated for targets such as red blood cells [1] and nanoparticles [2]. Those studies show that OT can trap and manipulate small objects in free flowing channels in relatively shallow tissue (50 μm) without any correction to the incoming beam.

The biological system that we aim to manipulate here is larger (50–60 μm in diameter) and is located much deeper into brain tissue (about 150 μm under the skin). To my knowledge, this represents one of the most challenging in-vivo optical trapping undertaken in combination with the size of the trapped object and the depth at which the trapping was performed.

In Chap. 5, I determined the optimal regions and positions where the optical traps should be positioned for maximal lateral forces on otoliths. In this chapter, we used those calculations and aimed to apply physical trapping forces to the two utricular otolith simultaneously in-vivo in zebrafish, observe behavioural response to these manipulations in different directions, and finally image and study the brain activity through which the fictive vestibular information is processed.

6.1 Experimental Set-Up

To be able to manipulate the two utricular otolith simultaneously, we need to create two optical traps and control them independently. From the literature, we can find a wide range of different methods developed to trap many particles at the same time. However, trapping multiple particles simultaneously and independently is a harder problem. Researchers found different ways to trap and manipulate multiple objects simultaneously [3]: with SLMs [4, 5], by scanning a single trap in 3D [6], and with dual-trap [7, 8], to cite the most interesting ones for our application.

© Springer International Publishing AG, part of Springer Nature 2018 63
I. A. Favre-Bulle, *Imaging, Manipulation and Optogenetics in Zebrafish*,
Springer Theses, https://doi.org/10.1007/978-3-319-96250-4_6

Also, to visualize the larva's behavioural responses, we need to image its whole body, because we are interested in the tail and eyes, since these are likely to be moved to compensate for perceived motion [9–11]. The need to apply two traps and image the body simultaneously led us to build a new set-up different from the set-ups presented previously, but with some similarities to the optogenetics system presented in Sect. 3.1.2. A sketch of the set-up is presented in Fig. 6.1.

Among the methods used to create multiple independent traps, the one which seemed to be the most flexible was the generation of multiple traps through the use of an SLM as discussed in Chap. 3. SLMs are widely used in optical trapping as they are very flexible, and can correct for aberrations [12] or sample disturbances [13, 14]. Moreover, pre-existing programs to drive an SLM are available online as previously discussed in Sect. 3.1.2. However, after building an optical trapping system with an SLM (very similar to the set-up presented in Fig. 3.3) we noticed that the amount of light present in each trap in the sample was far too low and could not provide enough force to manipulate the otoliths.

As light efficiency is an important factor, we decided to go for a different method which optimizes the transmission of light for the dual-trap OT system. A detailed sketch of the system is presented in Fig. 6.1a.

The dual OT system is composed of a infra-red (IR) laser (1070 nm IPG Photonics YLD-5 fibre laser), a half wave plate that rotates the laser polarisation by 45°, and a polarising beam splitter that splits the incoming beam into two beams of same intensity with perpendicular polarisation. Lenses L1 and L2 (100 mm focal length each), placed in each path of the two independent beams, are placed in a 4f config-uration (two focal length away from each other). The two beams are then reflected off of gimbal mirrors (GM) to be recombined with a second polarising beam split-ter. A second 4f configuration, comprising lenses L3 (100 mm focal length,) and L4 (200 mm focal length), increases the beam by a factor of two to fill the back aperture of the 20 × 1 NA Olympus microscope objective (XLUMPLFLN-W) after reflection on a 950 nm cutoff wavelength shortpass dichroic mirror in the imaging column. This optical system creates two diffraction limited spots at the imaging plane of the microscope objective. The positions (x, y) of each spot is dictated by the angle of each GM, and the third dimension (z) is determined by the distances between the lenses L1 and L2. In this study only (x, y) positions will be changed.

To precisely target the two traps on the otolith and image the eyes the same 20 × 1NA Olympus objective, tube lens L5 (180 mm focal length) and PCO edge 5.5 camera (Camera 1) were used. Finally, to image the whole larva and record its tail movements over time, a 4 × 0.1NA Olympus microscope objective (PLN 4×) was placed below the sample, and a tube lens L6 projected onto a Basler aca1920 camera (Camera 2) recording movements at 160 fps. Note that the 4× and 20× objectives are not collinear, the 4× being centred in the center of the fish body and the 20× in the center of the fish head.

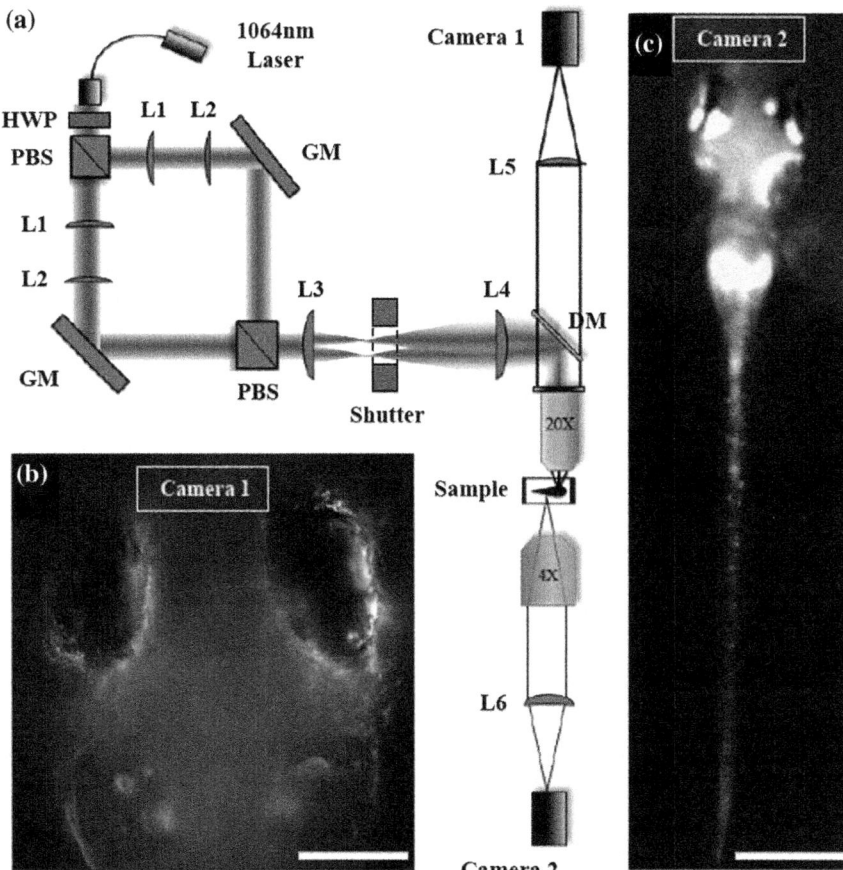

Fig. 6.1 Sketch of the optical trapping set-up for otolith manipulation and behavioural imaging. **a** Experimental set-up for delivering a dual OT to the larva using a 1064 nm fibre laser, a half wave plate (HWP), polarizing beam splitters (PBS), gimbal mirrors (GM), a dichroic mirror (DM), and lenses to collimate the beams and project the two traps into the sample chamber via a 20 × NA = 1 microscope objective. **b** Example of image recorded by Camera 1 which allows precise targeting of the otoliths with OTs and recording of the eyes motion (scale bar, 200 μm). **c** Example of image recorded with Camera 2 which permits the recording of tail movements (scale bar, 600 μm)

6.2 Measuring Behavioural Responses as a Result of Otolith Manipulation

As explained previously, the utricular otoliths are, to our knowledge, the most massive and most optically complex objects to be manipulated with OT. Whether OT will effectively apply forces in-vivo, where the otoliths are deep within a complex milieu of tissues, is uncertain. It is also not assured that the OT forces placed on otoliths will be sufficiently physiological to trigger vestibular behaviours in the larvae. To address these questions, we performed OT on otoliths and recorded only behavioural responses.

6.2.1   Experimental Procedures

Zebrafish larva 6 dpf were set in 2% LMA with the same experimental procedure presented previously in Sect. 3.2.1. The sample was placed under the 20× MO and the (x, y, z) position was adjusted with manual micrometer stages until the two urticular otoliths were in focus on Camera 1. The ensemble 4× MO, L6 and Camera 2 was translated up and down until the fish image on Camera 2 was at focus, this distance indeed slightly changes from fish to fish. Finally, the two focussed spots were manually moved in X and Y to precisely target the two utricular otoliths in pre-defined locations, we recorded eye motions at 20 Hz on Camera 1 and the tail at 160 Hz on Camera 2.

As discussed in the Chap. 5, we know that the strongest pulling forces happen when the trap are located few microns before reaching the edge of the otolith. By trying different directions multiple times in multiple zebrafish, I concluded that the fish were not responsive to pulling towards the anterior and posterior direction but were consistently responsive to pulling towards lateral and medial directions from very low laser power to more than 600 mW in each trap. The exposure time was also an important factor mainly due to heating. Very short exposures (100 ms) led to weak responses, and long exposures (3 s) drove chaotic escape responses, presumably due to heating of the tissue in the ear. These observations allowed us to refine our stimulus train to produce physiological responses in the animal that can be interpreted as vestibular behaviour.

Tail deflection was recorded for 5 different powers: 50, 100, 200, 400, and 600 mW. Those trials were targeting the lateral side of one otolith. Finally a 600 mW OT was placed in the centre of that same otolith to control for the possible effects of heating, as this treatment should produce equal heat without applying a coherent trapping force. Each trial was repeated 3 times consecutively with a 1 s exposure and 9 s delay between exposures. The different powers were presented to the animal in a random order. After those trials, the tail deflection was studied with 3 different trapping conditions: 600 mW OT on the lateral side of one otolith, 600 mW on the medial side of the other otolith, and dual traps (outside of one otolith and inside of the other, to produce a coherent fictive acceleration across both). Each trial was repeated 3 times with the same exposure time and waiting time between trials. The different combinations of OTs were presented to the animal in random order.

Every trapping condition was presented in three separate trials, and to ensure that spontaneous off-target behaviours were not being included in our data, we excluded trials in which:

- Spontaneous swimming occurred less than one second before OT onset,

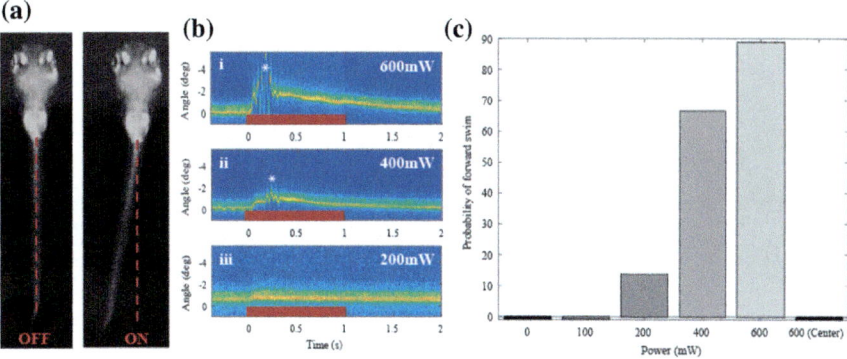

Fig. 6.2 Tail responses to lateral otolith traps. **a** Tail position of a larva before (left) and during (right) a 600 mW optical trap to the outside of the right otolith. **b** Tail position over time during trapping at different laser powers. Red bar marks when the trap is on. Forward swimming can be observed during the 400 and 600 mW trials (asterisk). **c** The probability of forward swimming increases with increased laser power (n = 5 larvae, 3 trials at each power). No forward swim has ever been observed for 600 mW of trapping power in the centre of the otolith

- Spontaneous swimming occurred less than one second after OT offset,
- Escape behaviour occurred during or within one second of the OT.

Over the three trials of every trapping condition, either two or three passed the exclusion tests and contributed to the data presented in Figs. 6.2, 6.3, 6.4, 6.5 and 6.6.

6.2.2 Behavioural Responses to Otolith Trapping

The results from these experiments show clearly and consistently that the targeting the lateral edge of one utricular otolith results in the deflection of the tail in the contralateral direction. Figure 6.2a shows snapshots of the lateral right otolith OT at 600 mW. The tail deflection is clearly visible.

To have a quantitative measurement of the tail deflection for each fish, I wrote a code in Matlab to determine and track horizontally (perpendicularly to the tail) the position of one visible pigment patch on the tail end. Finally the angle of deflection was calculated with simple trigonometry by setting the base of the tail as a vertex, the initial position of the tracked marker as one point, and the moving marker as the second point. Some examples of the data obtained from this calculation are presented in Fig. 6.2b. Not only the angle of deflection is calculated but the type of behaviour can be deduced from the responsive curves. Indeed, at the higher powers of 400 and 600 mW, a forward swimming motion (asterisks) happening just after the onset of the trap. This forward swimming occurs in most of the fish at 600 mW, with a decreasing probability with decreasing power (Fig. 6.2c).

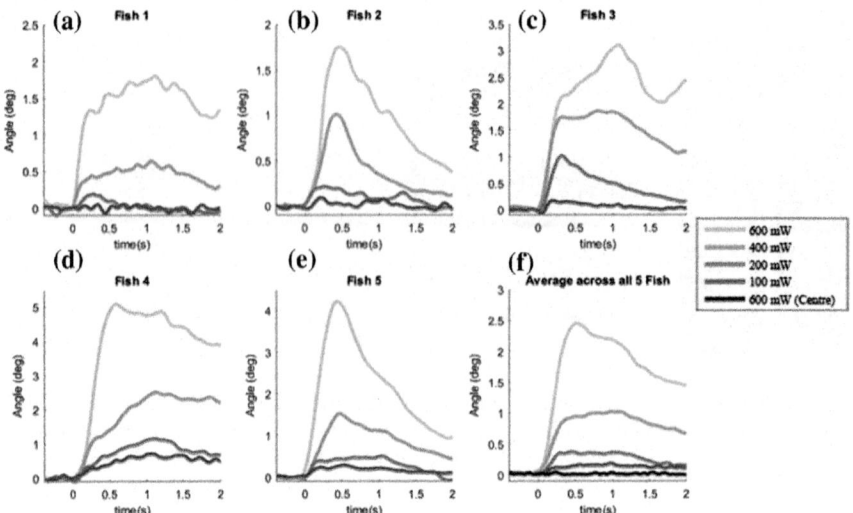

Fig. 6.3 Tail responses to otolith manipulation with laser power for 5 fish, and their average responses. Each curve represents the average of the 3 trials at one power. The tail deflection increases with laser power for each fish, which gives an average response across all 5 fish, also increasing with laser power. A 600 mW trap to the centre of the otolith does not trigger any tail motion

In each of those curves, an intensity threshold was set to filter out the position of the chosen pigment patch on the tail where the pixels intensities are the highest in the selected region. In the final step, the smooth function was used on the filtered data with the RLOESS method, a robust local regression using weighted linear least squares and a second degree polynomial model. Finally, the averages of the 3 trials with the same experimental conditions were calculated and single data point curves were obtained as shown in Fig. 6.3.

From Fig. 6.3 we can clearly see that, for the 5 fish presented, higher power OT resulted in increased angle of the tail deflection, and strong OT caused strong tail bend and oscillations representing a forward swimming motion (Fig. 6.2b and c). These responses were not caused by heating or pain because an OT spot targeted to the centre of the otolith (where there is no or negligible force) had no behavioural effect (Fig. 6.3). These results show that OT can provide a fictive acceleration stimulus that feeds into behaviourally relevant circuitry.

I next analysed the contributions that each of the two otoliths make to this behaviour. As demonstrated previously, when the OT is positioned on the lateral side of one otolith, which is equivalent to an outward force, it results in a contralateral bend of the tail on the trap onset. After the trap is turned off the tail gradually returns to its initial position (Figs. 6.2 and 6.3). From Fig. 6.4 we can see that an inward force on the opposite otolith had no effect at the time of OT onset, but that the tail makes a bend in the opposite direction on the trap offset (Fig. 6.4a ii and b yellow curves). Finally, trapping both otoliths simultaneously results in an active bend on the trap onset and active return to the baseline position upon offset (Fig. 6.4a ii and b green curves).

From the results presented in Fig. 6.4, we can conclude that the overall tail deflection response to a bilateral stimulus appeared to be the linear combination of the two otoliths' independent contributions, as the analysis of the responses across five larvae is consistent.

Since vestibular stimulation leads to compensatory eye movements [9], I also tracked the positions of the eyes during optical trapping (Figs. 6.5 and 6.6).

Eye motions were tracked in the same plane (from Camera 1, Fig. 6.1) as the otoliths targeting plane. Pigments on the retina were used as visible landmarks and were tracked (Fig. 6.5a) in a similar way as the tail pigments; their displacement were measured by recording their centre points overtime. The same Matlab code used for the tail was similarly used for the eyes. Rolling angles were calculated with simple trigonometry, using an average of 300 μm for the eyes height in 6dpf embryo, and assuming rotation around the centre of the eye. The laser power and exposure time were the same during eye tracking as they were for tail tracking.

In Figs. 6.5 and 6.6, the result of the lateral OT of the right utricular otolith shows that an outward force led to the simultaneous rolling of both eyes (Fig. 6.5b and c). As with the tail, these eyes movements were recorded in 5 different fish. The eyes moved more in response to stronger traps (Fig. 6.6a). Inward forces did not affect the eyes, even in trials where they drove resetting movements in the tail (data not shown).

6.2.3 Tail and Eyes Movements in the Literature

The consistent behavioural responses to otolith OT firmly establish that movements of the utricular otoliths alone are sufficient to drive compensatory responses across the body of a larval zebrafish.

From the literature, we know that the tail movements obtained here are similar to those resulting from optogenetic stimulation of neurons in a specific region of the zebrafish brain: the medial longitudinal fasciculus (nMLF) [15]. In this study, the activation of nMLF neurons led to ipsilateral deflections of the tail, with a return to the centre after the activation was stopped. Similarly, forward swimming was observed in response to powerful stimulation. These findings suggest that the nMLF may be involved in the transmission of the vestibular signals and information to the tail muscles.

Moreover, our observations of the eyes represent the vestibulo-ocular reflex (VOR), a three-neuron circuit that has been previously described in various vertebrates, including larval zebrafish [11]. In this study, changing a larva's pitch resulted in compensatory eye movements, while we see rolling eye movements in response to a perceived roll of the larva's body. This emphasizes the flexibility of the VOR in larval zebrafish, despite the simplicity of the larval nervous system.

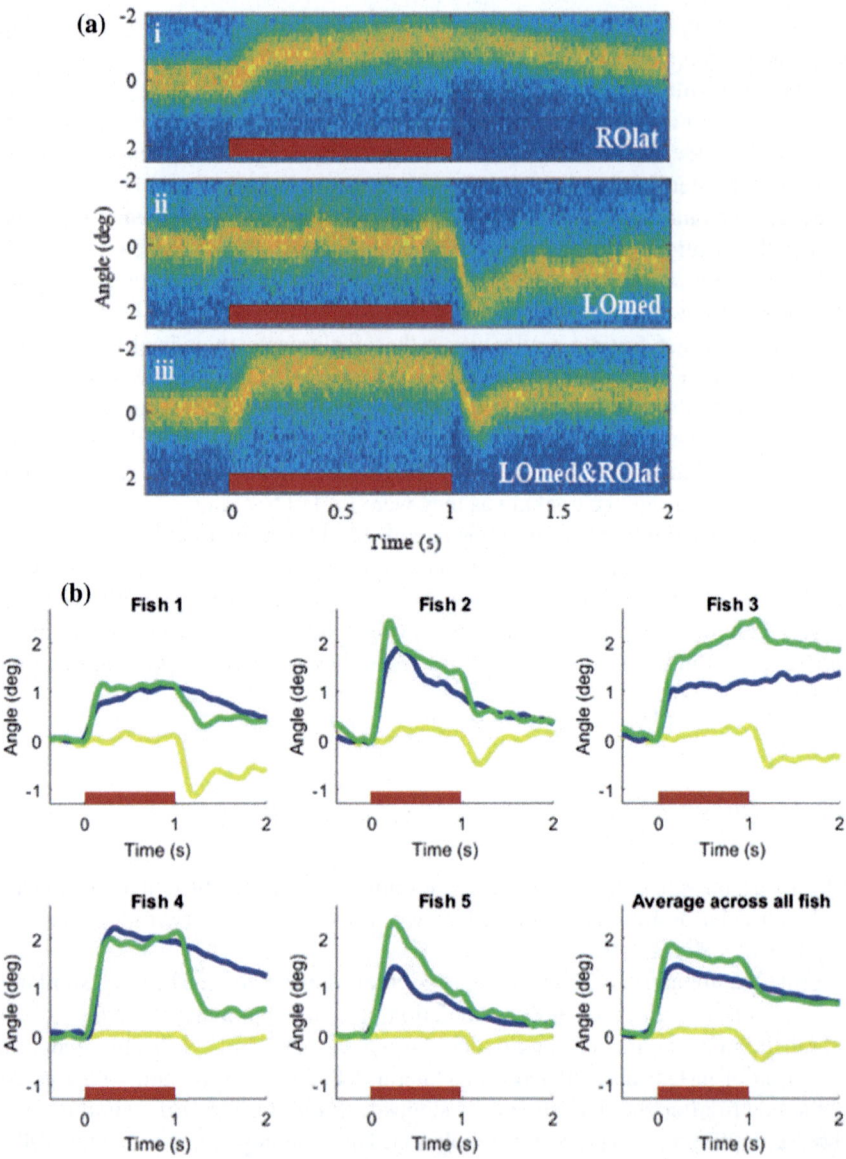

Fig. 6.4 Tail responses to lateral and medial otolith manipulation. **a i** Tail response to lateral right otolith (ROlat) trapping showing a left tail deflection on the trap onset and slow return on the offset of the trap. **ii** Tail response to medial left otolith (LOmed) trapping showing no response during trapping but a tail bend to the right on the trap offset. **iii** Tail response to ROlat and LOmed trapping showing deflection on the traps onset and active return on the traps offset. **b** Tail responses to ROlat (blue curves), LOmed (yellow curves) and both simultaneously (green curves) for 5 different fish and the average across all 5 fish. Red bars mark when the trap is on

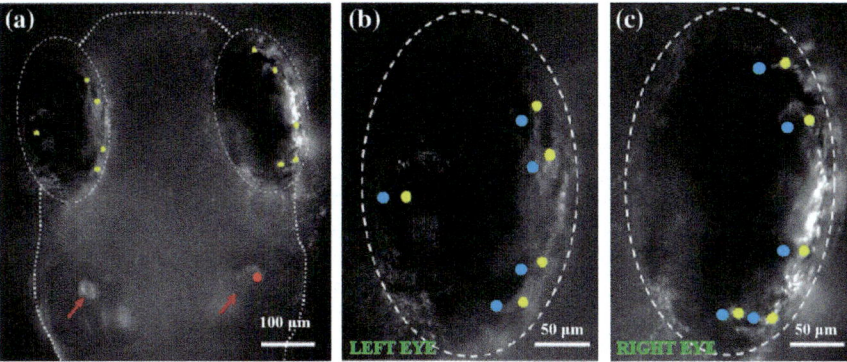

Fig. 6.5 Eye responses to otolith manipulation. **a** Identifiable retinal pigment granules (yellow dots) and the utricular otoliths (red arrows) are indicated. The site of the trap, on the lateral side of the right otolith is represented with a red dot. **b** The displacement of the tracked pigments (blue dots) is measured during the application of the trap. The original positions are indicated by yellow dots. In this example the lateral trapping of the right otolith led to simultaneous rolling of the eyes in the same direction

Those past studies and the neuronal circuitry described provide clues as to the pathway that this vestibular information takes through the brain. Exploring these pathways is the focus of the next section.

6.3 Set Up Improvements for Calcium Imaging

The next questions I want to address are the implications of behavioural observations on brain circuitry: is the information from both otoliths linearly processed in the brain, how are higher intensity traps encoded, and how is forward swimming encoded?

To visualize calcium activity over time with different trapping conditions as set in the previous section, further upgrades needed to be implemented on the set up presented in Fig. 6.1:

- First, we need to express GECIs markers in zebrafish brain. I will use GCamp6f as for the previous studies (Sects. 4.1 and 4.4.1). As the laser exposure lasts one second and we observe behavioural responses at onsets and offsets, it was important to separate those signals and have a fast marker such as GCamp6f.
- Second, I need an optical system to illuminate the brain regions I am interested in. We will use SPIMs (Fig. 6.7a) as shown for the optogenetics set up in Chap. 4. The SPIMs in a 90° angle can nicely image brain activity, and this is why we decided to implement this geometry here.

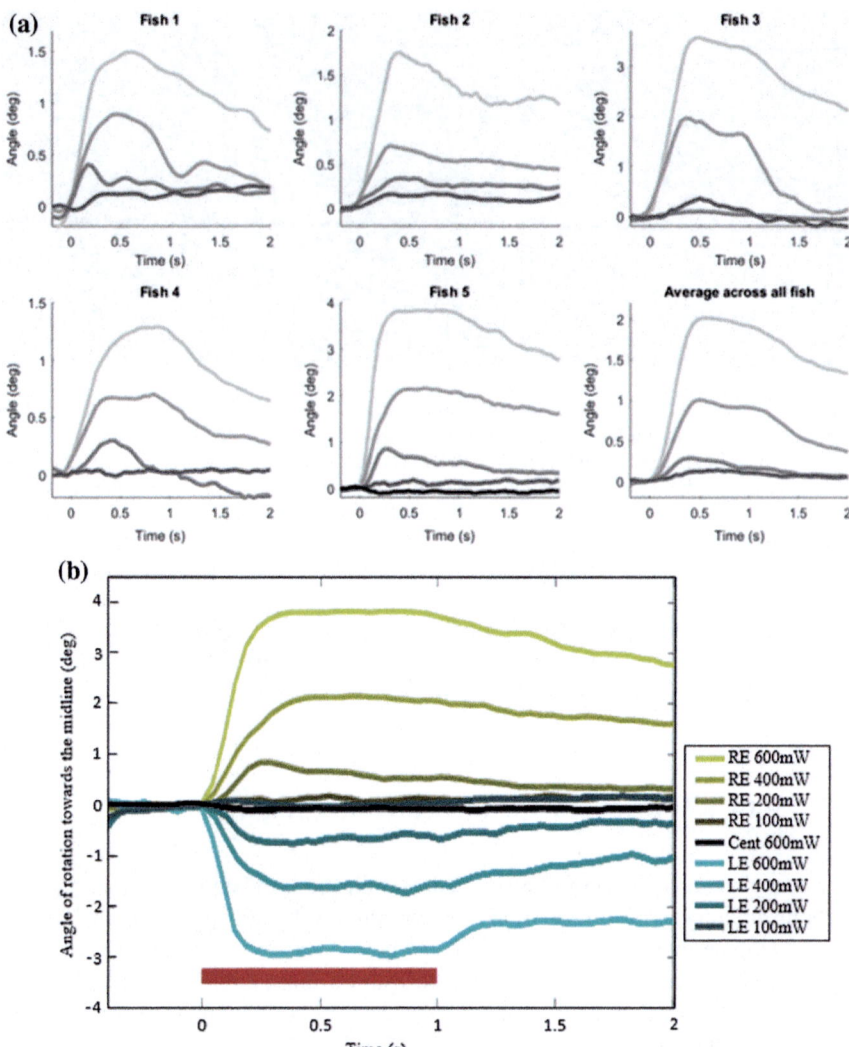

Fig. 6.6 Measurement of eye responses with power of otolith manipulation in 5 different fish. **a** Eye response curves for each of the 5 fish and the average across all 5 fish. OT powers are 600, 400, 200, and 100 mW (light to dark curves). **b** Rotation of eyes rolling over time. The left eye (LE, blue) and right eye (RE, yellow) rotations in the direction towards the midline with the same 5 powers than in a. Since the eyes roll in the same direction, this results in opposite values for the left and right eyes with reference to the midline. Each curve is the average of three trials with same experimental conditions

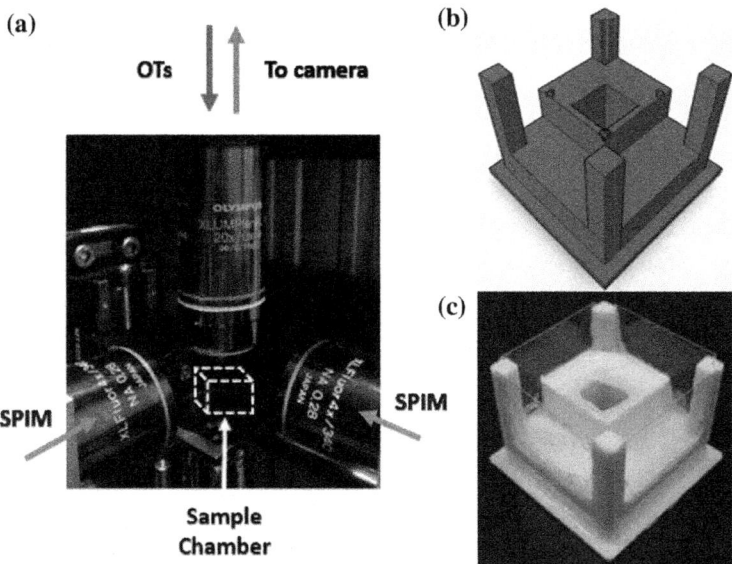

Fig. 6.7 Set up improvements for calcium imaging. **a** Two perpendicular SPIMs are positioned perpendicular to the imaging system. **b** Sample chamber designed on Tinkercad. **c** Sample chamber 3D printed. 20 × 20 mm glass coverslips are glued to the four wall posts with transparent nail polish (non toxic for zebrafish). A 10 × 10 mm glass coverslip is glued on the bottom window for tail imaging

- Another important element to upgrade was the sample chamber: to avoid curved surfaces along the SPIMs path, chambers have been designed on Tinkercad as described in Sect. 4.4.1. However, as I need to image the fish from below, a window under the fish needed to be created, so a new chamber has been designed and 3D printed (Fig. 6.7b and c).
- Lastly, I need to image brain activity at multiple depth. In Chap. 4, we used a motorised stage to move the sample up and down to target the different depth of interest. In this system I had to move the optical traps up and down to stay in the otolith plane. From Fig. 6.1 we know that the Z parameter can be changed manually with lenses L1 and L2. These manipulations cause x- and y-drifts as z changes. One solution is to use an Electrically Tunable Lens (ETL), a lens in which focal length can be modified electrically. The ETL will be used similarly to the system described by Fahrbach et al. [16]. This implies that the sample won't move, but only the imaging plane will (Fig. 6.8). As the ETL will image different planes along Z axis, the two SPIM planes will also need to be translated along Z axis to be positioned at the variable imaging plane heights dictated by the ETL. Manual micrometer stages are used to follow the translation of the imaging plane.

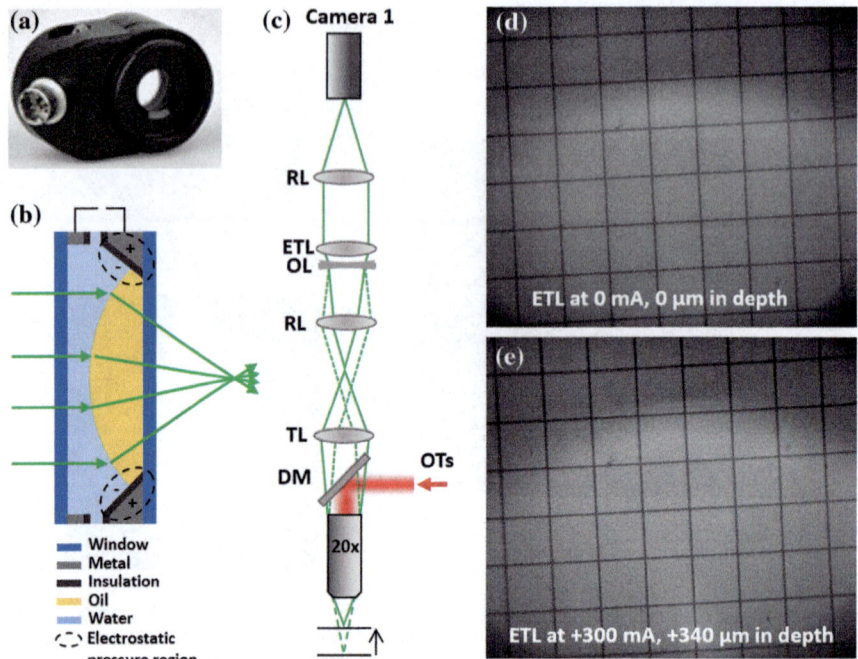

Fig. 6.8 ETL functioning and implementation in an optical system. **a** Picture of the ETL from Optotune EL-10-30-C-VIS-LD-MV with an in-built Offset lens (OL). **b** Sketch of light diffraction in an ETL when current is ON. Incoming current creates an electrostatic pressure compressing the oil and resulting in a curved surface and short focal length of the resulting lens. **c** Sketch of the implementation of the ETL in our imaging system. Light from the sample is collected by the MO, transmitted to its Tube Lens (TL) of 180 mm. The BFP of the MO is then imaged on the ETL via a Relay lens (RL). A second relay lens transmits the imaging plane onto Camera 1. **d** Example of a 100 μm grid imaged when no current runs through the ETL. **e** The same grid moved +340 μm in Z and imaged on the camera when the ETL current is to its maximum 300 mA. No obvious distortions are visible and the scale is the same

6.3.1   Details of ETL Functionning and Performances

Electrically Tunable Lenses (ETL) (Fig. 6.8a) are glass chambers containing water and oil (see details in Fig. 6.8b). The electrodes connected to this chamber create an electrostatic pressure on the oil which changes its curvature. With stronger currents, the oil surface curvature increases which results in a shorter focal length lens. Working with water and oil also implies that the tunable lens need to lie horizontally in the set up and that the room temperature or device temperature has an effect on the curvature. When using the stronger currents allowed by the device (300 mA), I indeed noticed that the device temperature, originally at room temperature (around 24°) was rising up to 30° and a few minutes were needed for the imaging plane to stabilize. This feature was taken into account in the recordings.

The imaging set-up sketched in Fig. 6.8d allows the imaging of multiple planes on Camera 1 with magnifications and travel distances described in Fahrbach et al. paper [16]. Being aware of the compromise to be made between the travel distance (longer with longer RL focal length) and the loss of information due to the small ETL clear aperture of 10 mm (more loss with longer RL focal length), we set the RL to be 120 mm allowing the travel distance in Z to be about 340 μm which covers the whole the brain for 6dpf fish and the loss of information to be negligible.

Originally the ETL comes with an in-built Offset Lens (OL) of -150 mm which allows for our system to scan 170 μm above and 170 μm below the 20$\times$ MO focal plane, where the optical traps focus and utricular otoliths are located. As the brain is mostly above the utricular otolith, half of the scanning range is lost. To avoid this drawback, the OL was replaced with a -300 mm lens which allows the ETL to scan 340 μm above the MO focal plane.

Figure 6.8 show images recorded on Camera 1 when the ETL is at rest (0 mA) and set at its maximum current (300 mA). No distortions are visible. The scale increase was calculated to be of about 1.3%, which is negligible for the most extreme planes. These results show that different planes in the brain and neurons over the brain will be imaged with similar optical conditions.

6.4 Calcium Imaging: Mapping Out Responses to Acceleration

Calcium imaging is a well established method to image active neurons over time and understand brain circuitry. In the context of this study, I want to find which brain regions are involved in the utricular otolith movement/acceleration in zebrafish.

6.4.1 Past Relevant Studies on Brain Activity in Zebrafish

The zebrafish vestibular system has, to our knowledge, not been mapped out, but a few studies provide clues on where we should expect activity.

Tail deflection in zebrafish has been observed in past studies [15] using optogenetics and illuminating a very specific region of the brain called nucleus of the medial longitudinal fasciculus (nMLF). This brain region located deep in the brain (see Fig. 6.9) is constituted of few tens of neurons on each side of the brain. The Thiele et al. study proves that illuminating or triggering activity on one side would elicit a tail bend in the ipsilateral direction. This result suggests that the nMLF is involved in the control of the tail and is potentially involved in the behavioural response to acceleration observed.

Another aspect of optical trapping we need to be aware of is heat. The laser powers I use are high and create heat within the fish. Even though the behavioural

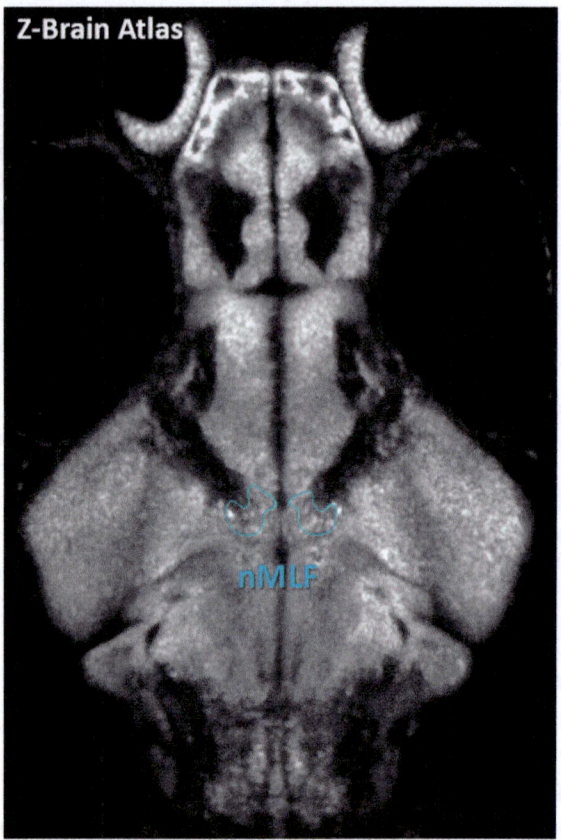

Fig. 6.9 nMLF location in Zebrafish brain using Z-Brain atlas on http://engertlab.fas.harvard.edu: 4001/

responses observed and shown cannot be from heat, it does not prove the absence of heat sensation and heat information processing in the brain. Noxious heat has been studied in zebrafish [17–19], and multiple brain regions show activity in response to noxious heat. With those results, we know that we need to differentiate brain activity in responses to heat and acceleration. The solution is to define a slightly different experimental procedure.

6.4.2   Experimental Procedure

Zebrafish larva 6dpf was embedded in 2% LMA on a 10×10 mm coverslip. This coverslip was placed on the bottom window of the sample chamber (shown in Fig. 6.7) with a drop of 2% LMA to ensure stability. The chamber was placed under the microscope and the utricular otoliths were positioned at camera focus when the ETL was at rest. The OTs were targeted to the otoliths, the two SPIM planes were turned

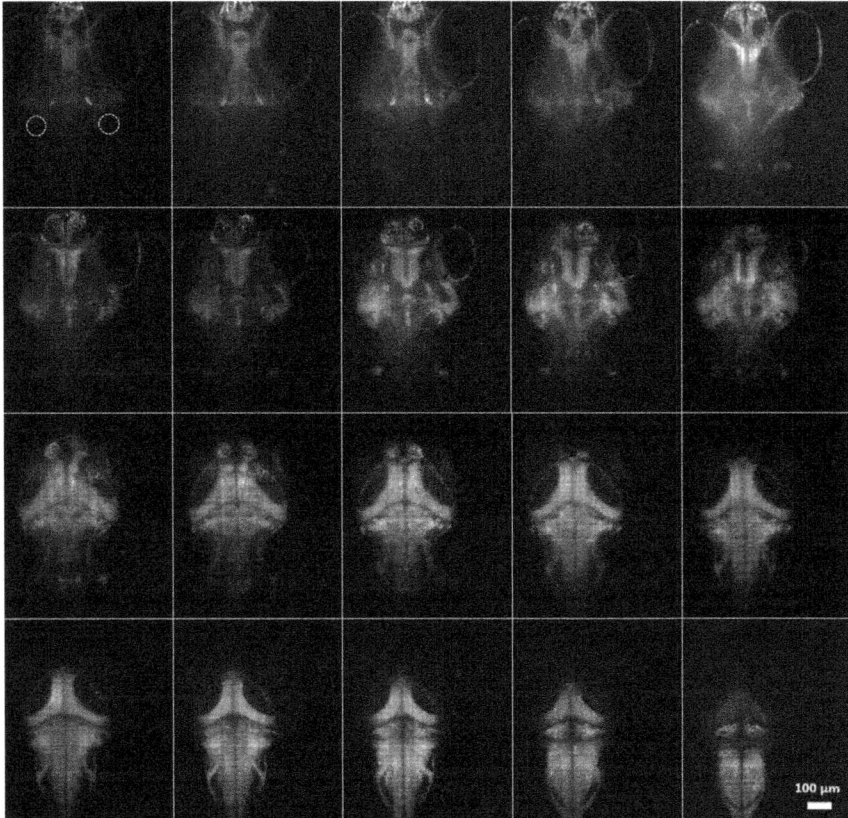

Fig. 6.10 Multiple planes imaged with varying ETL focus. Top left: MO focal plane is imaged when ETL is at rest, utricular otoliths are circled in red. Bottom right: Shallower plane imaged, +260 μm above the focal plane. Planes were imaged at 13 μm increments

on and positioned in the utricular otolith planes. An example of images obtained at that depth is presented in Fig. 6.10.

Trapping at the lateral and center position of the otolith were tested to distinguish the neuronal activity coming from heat and acceleration sensing. Each trial was repeated 3 times consecutively with a 1 s exposure at 400 mW and 9 s delay between exposures as for Sect. 6.2.1. The best location of the trap on the lateral side was determined by the position which would elicit the highest angle roll of the eyes. No motions of the eyes were detected when the trap was positioned in the otolith's centre. Once one brain plane was imaged, the OTs and fish remained stationary while the ETL and SPIM planes were adjusted to a different focal depth. This new plane was imaged under the same OTs conditions and so on.

6.4.3 Preliminary Results

To have an overall view of brain activity and the regions involved in the acceleration processing, the whole brain was imaged with $13\,\mu$m steps starting from
the otoliths' plane and finishing at the shallowest plane possible under the skin.
Figure 6.10 presents an example of the planes imaged in one fish. Typically, we have
about 20 planes, but this number can vary from fish to fish, due to the variations in
size and mounting orientation.

Once these data were recorded, I next need to identify consistently responsive
neurons, along with their locations in the brain. However analysing the videos to:

- identify neurons within a plane,
- differentiate the activity of a neuron in the plane from potential overlapping neurons,
- determine the spike train of each neuron from its fluorescent intensity profile.

is very challenging and complex. Past studies have approached separately those
studies [20–23]. Very few methods cover these three problems but, Pnevmatikakis
et al. [24] is such a method and has the advantages of dealing with big datasets by
running computations in the cloud, giving clean results by discarding artefacts. It
also is one of the top performers in the neurofinder competition.

Prior to the analysis, the videos need to be preprocessed. First, to avoid artefacts
due to eyes motion and light from the OT but also to have faster analysis, the original
videos need to be cropped to remove the eyes, ears and unnecessary pixels. Second, a
motion correction macro is run on all the videos to correct for any motion of the fish
during the experiments. This macro, developed by the NeuroTechnology Center at
Columbia University is available online at https://github.com/NTCColumbia/moco.

To be able to work on big data sets (one fish comprises 40 videos and about
30 GB of data) and analyse videos in a reasonable amount of time the videos are
loaded on a virtual machine on the Queensland Research and Innovation Services
Cloud (QRIScloud). From this platform, all the calculations and analysis can be done.
The computational toolbox developed by Pnevmatikakis et al. which we use for the
analysis of our large dataset, is called Calcium Imaging Analysis (CaImAn) and is
available on the following website: https://github.com/simonsfoundation/CaImAn.

Briefly, the toolbox first initialise regions of interest (ROI) using spatial gaussian filtering followed by non-negative matrix factorisation (NMF) to differentiate
ROIs signals. This is an iterative process to separate signals which are mixed due to
potential overlapping neurons and signals. When the correlation between overlapping
ROIs is too strong, the ROIs are merged. Finally, the toolbox extracts components
(or intensity traces) corresponding to each individual neurons.

Examples of data obtained from one fish with this analysis are presented in
Figs. 6.11, 6.12 and 6.14.

As a reminder, each video recorded 3 consecutive activations of the trap positioned
in the center of the otolith or in the lateral edge with a 1 s exposure time at 400 mW
and 9 s delay between exposures. The videos start one second before the first exposure

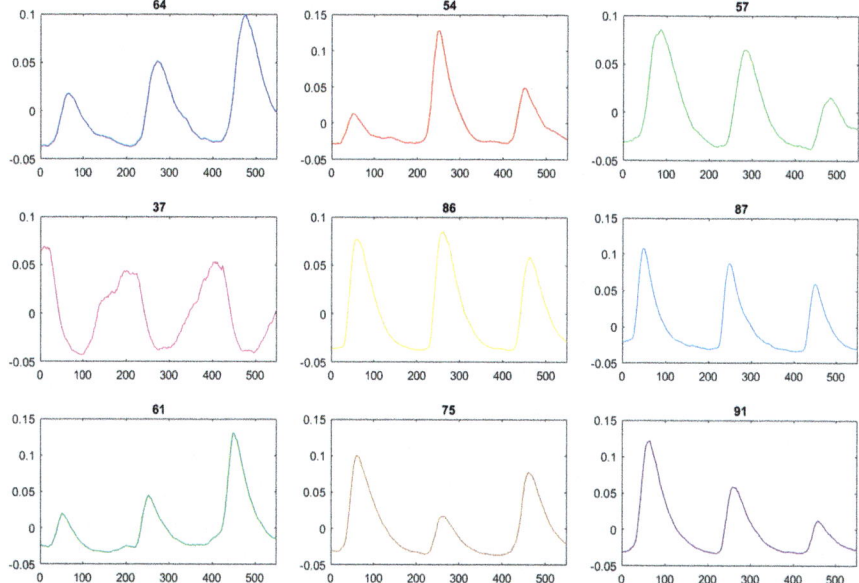

Fig. 6.11 Components selected of responsive neurons for one fish. Each component profile is represented overtime (540 frames with 20 fps). Trap exposures last 20 frames and appear at frames 20, 220 and 420. The number on top of each panel represent the number of neurons belonging to the component represented below. All the components are responsive to the three trap exposures

and end 5 s after the last trap. The videos are recorded at 20fps, giving a total of 540 frames.

For this particular fish, 9 components have been found for the two trapping conditions and each component show responses to each of the three exposures at frames 20, 220 and 420. Moreover, each component found represents at least 37 neurons which makes them strong representations of neuronal activity. In total, 612 neurons have been found responsive to the traps. Figure 6.11 shows plots of the intensity profile of each component over time, as selected by the code (the details of each neuron trace is represented in Fig. 6.12 second column). Only one component represent the silencing of neurons (pink curve), as the intensity of those neurons decreases when the trap is turned on. The 8 other components represent an increase of fluorescence for each exposure with peaks increasing, decreasing, or constant with the number of exposure.

Figure 6.12 gives more information on the 9 components selected. It represents the intensity traces of each neuron belonging to each component (second column). Visually, we can see that each selected neuron is strongly correlated to the component it has been assigned to. Figure 6.12 shows also the distribution in Z of the neurons belonging to each component. The distribution is quite dispersed for the 8 components which implies that no particular brain regions respond differently to the trap exposure. Only the silencing component appears to be more localised and deep

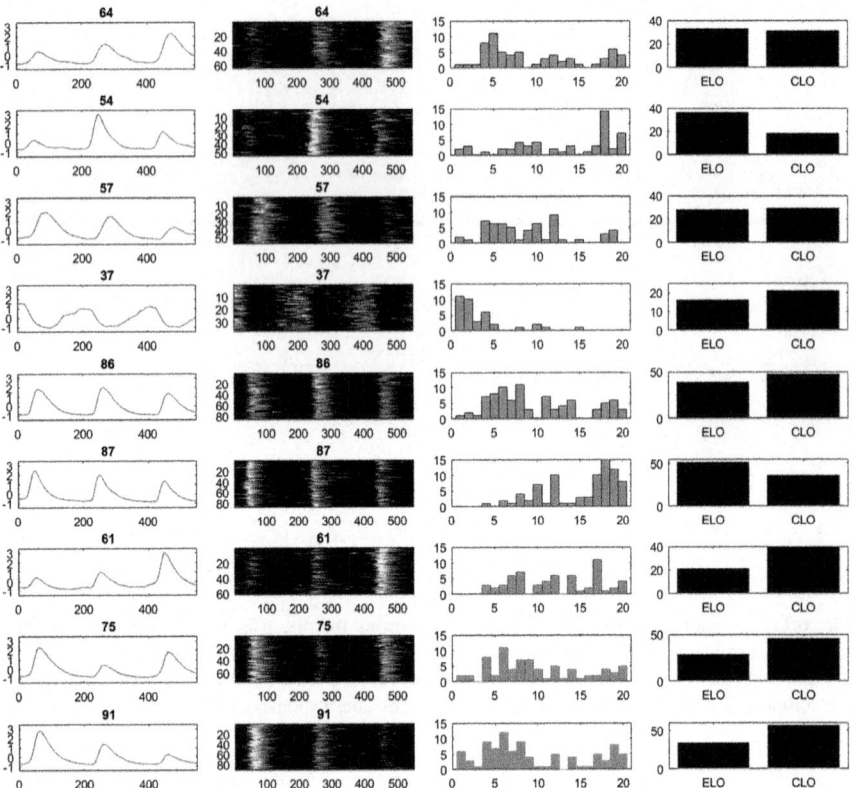

Fig. 6.12 Components, neuronal activity traces and location for one fish. First column represents the 9 components overtime as described in Fig. 6.11. Second column represents all the intensity traces overtime of the neurons from each component. Third column represents the distribution of location of the neurons from each component across the 20 planes recorded. Fourth column represents the distribution of the neurons with trapping modalities: exterior left otolith (ELO) and center left otolith (CLO) trapping

in the brain in this fish. Lastly, the fourth column of Fig. 6.12 shows for which type of trap elicits responses from the neurons in the component. Ideally we would have some components that are largely present for one type of trapping and not the other, however we find that all components respond comparably to both types of trap.

This analysis has been done for 4 different fish and is presented in Fig. 6.13. The 8 components found for those 4 fish, which represent 2326 neurons responsive to trap exposures, are scattered all over the brain and are not specific to one of the trapping modalities. These results do not help in distinguishing between acceleration and heat sensing and further analysis need to be done.

To further investigate the location of the responsive neurons in space and distinguish between the two trapping modalities, Fig. 6.14 shows in colors the location of

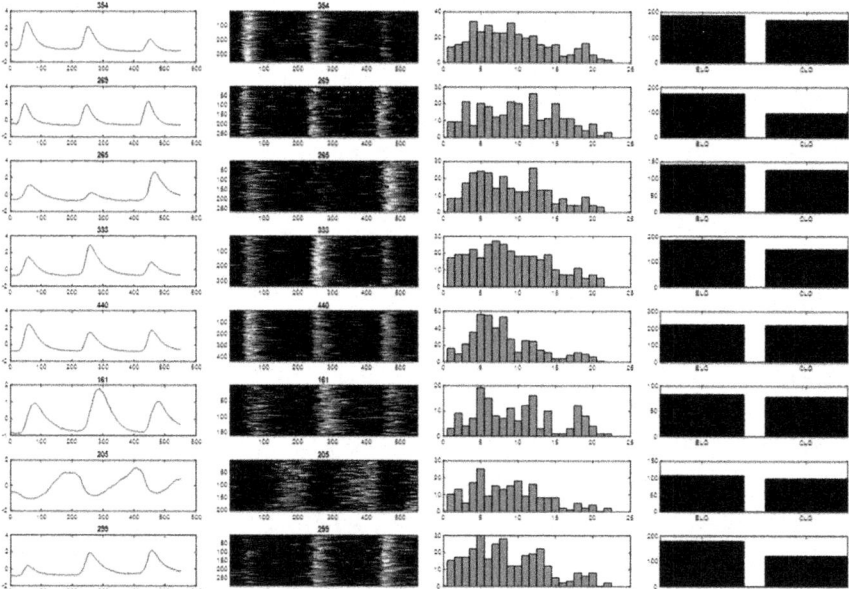

Fig. 6.13 Components, neuronal activity traces, and location for four fish. First column represents the 9 components over time as described in Fig. 6.11. Second column represents all the intensity traces over time of the neurons from each component. Third column represents the distribution of location of the neurons from each component across the 20 planes recorded. Fourth column represents the distribution of the neurons with trapping modalities: exterior left otolith (ELO) and center left otolith (CLO) trapping

the neurons detected. Their colors represent the component they belong to from the color code chosen in Fig. 6.11. For clarity 7 planes over the 20 are represented.

The subpallium (round dotted line), thalamus (dashed dotted line), torus semicircularis (solid lines), Medial Octavolateralis Nuclei (dashed line) are regions that respond to trap exposure in this fish, but have also been responsive in all the 4 fish analysed.

As discussed before, the subpallium is believed to be responsive to noxious heat [17], which explains why it is responsive for both trapping modalities. The thalamus is interestingly responding when trapping the otolith's edge and not responsive when trapping its centre, which could suggest that it is involved in the processing of acceleration information. The torus semicircularis and medial octavolateralis nuclei are clearly responsive to both trapping modalities however they are believed to be involved in motion detection [25–27] and not heat. This last result implies that, despite trapping the centre of the otolith and not seeing any eye motions, the otolith could undergo a weak but sufficient force to be detected and processed. We know from Sect. 5.3.2 that otolith are inhomogeneous in the centre region; cracks are visible and the force measured is very irregular in terms of magnitude and direction. These irregularities could indeed be sufficient to create a force perceived by the hair

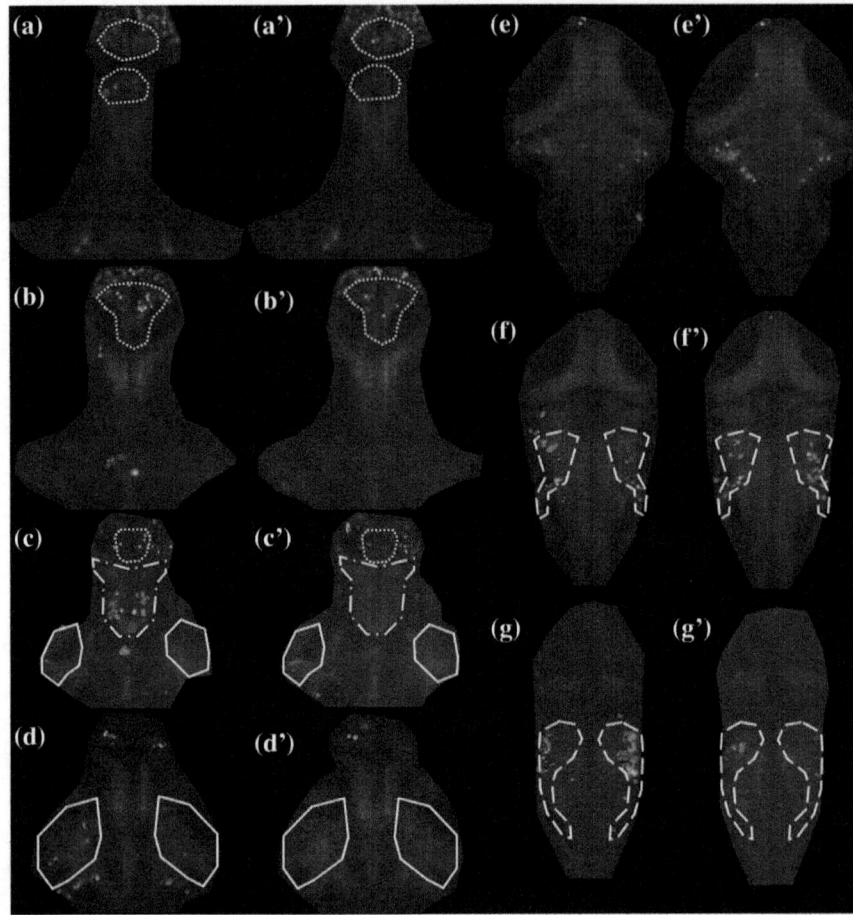

Fig. 6.14 Location of responsive neurons across the brain. The colors of the responsive neurons correspond to the color of the component they belong to and presented in Fig. 6.11. **a–g**. Location of responsive neurons with increasing distances from the otolith plane in response to lateral trapping. **a'–g'**. Location of responsive neurons with increasing distances from the otolith plane (same planes that a–g) in response to centre trapping. Round dot lines delimits subpallium, dash dot line the thalamus, solid lines the torus semicircularis, dash line the medial octavolateralis nuclei

cells. Moreover, the trap can also be slightly offset from the otolith centre along the Z axis which would elicit an additional force along the Z axis and contribute so an acceleration information.

These last results suggest that a new optical trap exposure modality needs to be determined such that it does not activate brain regions that we know are responsive to motion. One solution would be to place the trap in the ear liquid, away from the otolith, so that there is no interaction between the otolith and the trap but the heat is still created in the ear. These data are currently in the process of being gathered and analysed.

6.5 Discussion

Here, I demonstrate the tractability of OT, even on large objects such as otolith deep in-vivo despite the challenge of light scattering in brain tissue. Using optical traps, sufficient forces could be applied in different directions to simulate acceleration and to look at the reaction of the fish depending on the amplitude of the force and its directions. Results show that the tail and the eyes are directly involved in the acceleration sensory system and that the responses to a bilateral stimulus appeared to be the linear combination of the two otolith's independent contributions.

An important advantage of OT in this motion-detecting system is to remove the physical movement and the non-invasive aspect of the experiments. This allows a stationary imaging platform in which we can easily record neuronal activity over time during experiments. Imaging the neuronal activity in response to OT revealed that multiple brain regions could be involved in the acceleration information processing, however heat effects are also present and the current results do not allow me to isolate acceleration information processing from noxious heat. Further investigations are currently being undertaken.

References

1. M.-C. Zhong, X.-B. Wei, J.-H. Zhou, Z.-Q. Wang, Y.-M. Li, Trapping red blood cells in living animals using optical tweezers. Nat. Commun. **4**, 1768 (2013)
2. P.L. Johansen, F. Fenaroli, L. Evensen, G. Griffiths, G. Koster, Optical micromanipulation of nanoparticles and cells inside living zebrafish. Nat. Commun. **7** (2016)
3. D.G. Grier, A revolution in optical manipulation. Nature **424**(6950), 810 (2003)
4. J.E. Curtis, B.A. Koss, D.G. Grier, Dynamic holographic optical tweezers. Opt. Commun. **207**(1–6), 169 (2002)
5. D. Preece, S. Keen, E. Botvinick, R. Bowman, M. Padgett, J. Leach, Independent polarisation control of multiple optical traps. Opt. Express **16**(20), 15897 (2008)
6. K. Visscher, G.J. Brakenhoff, J.J. Krol, Micromanipulation by multiple optical traps created by a single fast scanning trap integrated with the bilateral confocal scanning laser microscope. Cytometry **14**(2), 105 (1993)
7. C. Bustamante, Y.R. Chemla, J.R. Moffitt, High-resolution dual-trap optical tweezers with differential detection: instrument design. Cold Spring Harbor Protoc. **2009**(10), pdb.ip73 (2009)
8. S. Rancourt-Grenier, M.T. Wei, J.J. Bai, A. Chiou, P.P. Bareil, P.L. Duval, Y. Sheng, Dynamic deformation of red blood cell in dual-trap optical tweezers. Opt. Express **18**(10), 10462 (2010)
9. W. Mo, F. Chen, A. Nechiporuk, T. Nicolson, Quantification of vestibular-induced eye movements in zebrafish larvae. BMC Neurosci. **11**(1), 1 (2010)
10. A.H. Groneberg, U. Herget, S. Ryu, R.J. De Marco, Positive taxis and sustained responsiveness to water motions in larval zebrafish. Front. Neural Circ. **9**, 9 (2015)
11. I.H. Bianco, L.H. Ma, D. Schoppik, D.N. Robson, M.B. Orger, J.C. Beck, J.M. Li, A.F. Schier, F. Engert, R. Baker, The tangential nucleus controls a gravito-inertial vestibulo-ocular reflex. Curr. Biol. **22**(14), 1285 (2012)
12. K.D. Wulff, D.G. Cole, R.L. Clark, R. DiLeonardo, J. Leach, J. Cooper, G. Gibson, M.J. Padgett, Aberration correction in holographic optical tweezers. Opt. Express **14**(9), 4169 (2006)
13. H.I.C. Dalgarno, T. Cizmar, T. Vettenburg, J. Nylk, F.J. Gunn-Moore, K. Dholakia, Wavefront corrected light sheet microscopy in turbid media. Appl. Phys. Lett. **100**(19), 191108 (2012)

14. M. Nixon, O. Katz, E. Small, Y. Bromberg, A.A. Friesem, Y. Silberberg, N. Davidson, Real-time wavefront shaping through scattering media by all-optical feedback. Nat. Photonics **7**(11), 919 (2013)

15. T.R. Thiele, J.C. Donovan, H. Baier, Descending control of swim posture by a midbrain nucleus in zebrafish. Neuron **83**(3), 679 (2014)

16. F.O. Fahrbach, F.F. Voigt, B. Schmid, F. Helmchen, J. Huisken, Rapid 3D light-sheet microscopy with a tunable lens. Opt. Express **21**(18), 21010 (2013)

17. O. Randlett, C.L. Wee, E.A. Naumann, O. Nnaemeka, D. Schoppik, J.E. Fitzgerald, R. Portugues, A.M.B. Lacoste, C. Riegler, F. Engert, A.F. Schier, Whole-brain activity mapping onto a zebrafish brain atlas. Nat. Methods **12**(11), 1039 (2015)

18. V. Malafoglia, M. Colasanti, W. Raffaeli, D. Balciunas, A. Giordano, G. Bellipanni, Extreme thermal noxious stimuli induce pain responses in zebrafish larvae. J. Cell. Physiol. **229**(3), 300 (2014)

19. M. Haesemeyer, D.N. Robson, J.M. Li, A.F. Schier, F. Engert, The structure and timescales of heat perception in larval zebrafish. Cell Syst. **1**(5), 338 (2015)

20. S.L. Smith, M. Hausser, Parallel processing of visual space by neighboring neurons in mouse visual cortex. Nat. Neurosci. **13**(9), 1144 (2010)

21. K.D. Harris, R.Q. Quiroga, J. Freeman, S.L. Smith, Improving data quality in neuronal population recordings. Nat. Neurosci. **19**(9), 1165 (2016)

22. P. Kaifosh, J.D. Zaremba, N.B. Danielson, A. Losonczy, SIMA: Python software for analysis of dynamic fluorescence imaging data. Frontiers Neuroinformatics **8**, 80 (2014)

23. E.A. Mukamel, A. Nimmerjahn, M.J. Schnitzer, Automated analysis of cellular signals from large-scale calcium imaging data. Neuron **63**(6), 747 (2009)

24. E.A. Pnevmatikakis, D. Soudry, Y. Gao, T.A. Machado, J. Merel, D. Pfau, T. Reardon, Y. Mu, C. Lacefield, W. Yang, M. Ahrens, R. Bruno, T.M. Jessell, D.S. Peterka, R. Yuste, L. Paninski, Simultaneous denoising, deconvolution, and demixing of calcium imaging data. Neuron **89**(2), 285 (2016)

25. G.E. Meredith, A.B. Butler, Organization of eighth nerve afferent projections from individual endorgans of the inner ear in the teleost, astronotus ocellatus. J. Comp. Neurol. **220**(1), 44 (1983)

26. C.A. McCormick, The organization of the octavolateralis area in actinopterygian fishes: a new interpretation. J. Morphol. **171**(2), 159 (1982)

27. W. Plassmann, Sensory modality interdependence in the octaval system of an elasmobranch. Exp. Brain Res. **50**(2–3), 283 (1983)

Chapter 7
Conclusion

To conclude this thesis, I will summarise the multiple studies that were performed which comprise the modelling and measurement of light scattering in brain tissue, how we used those results for current and future studies, and finally how we used light and combined optical systems to overcome the invasive nature of pre-existing methods in the study of vestibular system in zebrafish.

The first study performed investigated the problem of light scattering in brain tissue theoretically and experimentally, as it is of great importance for optogenetics which uses light to drive and visualize neural activity. It is essential to be able to quantify and predict light scattering so that optogenetic illumination can be gauged accurately. To do so, I presented and justified the model of brain tissue chosen and the Monte Carlo method developed to predict light scattering of a single spot with depth. In this thesis, I chose to work with a transparent and dense brain region such as zebrafish, however, the model and simulations can be adapted to any brain structure or scattering parameters wanted. The simulation of the propagation and spreading of a focussed spot showed that we can restrict illumination to single targeted neurons with a focussed spot relatively deep in dense brain tissue. However, the important light spreading along Z axis implies that experiments should be conducted with a careful adjustment of laser power (determined by the light source).

To support this model, I collected experimental data in zebrafish larvae, as presented in Chap. 3. The back-scattered light from a single spot has a very specific profile which can be explained by the model as an infinite sum of Gaussians, each Gaussian coming from a single plane above or below the focal plane and being slightly deformed by scattering. The understanding and quantification of the effect of scattering on a single spot deep in tissue led us to conclude that the correction for complex 2D illumination of groups of neurons is not necessary. I also demonstrated the construction of a simpler optogenetics system.

As reported in the literature, a wide range of optical systems have been optimised to be part of optogenetics systems, however the specificity of zebrafish as a model led us to construct our own combination of optical systems with two SPIMs and

© Springer International Publishing AG, part of Springer Nature 2018
I. A. Favre-Bulle, *Imaging, Manipulation and Optogenetics in Zebrafish*,
Springer Theses, https://doi.org/10.1007/978-3-319-96250-4_7

an SLM. One example of the questions we can answer with this new system is presented in Chap. 4, where we illuminated multiple neurons in several regions in the brain and deduced the multi-sensory property of the brain region of the optic tectum. Furthermore, the flexibility of SLM allow multiple volumes illumination of multiple and single neurons sizes at various depth independently or simultaneously which can solve a wide variety of problems on brain circuitry in zebrafish. Moreover, the laser we used for ChR2 activation being of 488nm (low wavelength) suggests that the targeting performance we accomplished can be further improved with the use of longer wavelength lasers, since they have longer penetration depth and tissue appear clearer at those wavelengths. This technology could be used to drive or silence very specific neurons and provide to a potentially powerful tool for circuit analysis.

In the next study, I aimed to gain further understanding of light's interactions with biological systems and used light to investigate the optical properties and neuronal encoding of the ear-stones (or utricular otoliths) responsible for acceleration sensing. From the literature, we know that utricular otolith are made out of more than 95% of calcium carbonate crystal, which implies that they are transparent and have a refractive index greater that their surrounding (close to water). Those two properties should permit optical trapping. I tested the feasibility of this manipulation by extracting otoliths from the zebrafish and manipulating them free in water. We know that if we can apply sufficient force to simulate acceleration in a swimming fish, we needed be able to quantify and locate the maximum forces that can be applied. Optical forces were modelled using the ray optics method and direct measurements were taken using light deflection method. Results from the model and the experiments show that we can apply forces in the order of pN, which is what an otolith would encounter during typical swimming. As the model and measurements have been made for an otolith in water, these results give an upper limit of the force we can apply in-vivo, since biological tissue will introduce scattering and beam distortions.

In the final Chapter, I presented the study of behavioural responses and brain activity in response to utricular otolith manipulations. Results showed that the tail and the eyes are directly involved in vestibular processing, and that the responses to a bilateral stimulus appeared to be the linear combination of the two otolith's independent contributions. These results prove once more that optical traps provide sufficient forces to simulate acceleration in various directions. As OT are non-invasive, it makes it a very attractive new method for the study of vestibular system in zebrafish. Removing the physical movement, seen in pre-existing methods, allows a stationary imaging platform where neuronal activity can easily be studied.

Imaging the neuronal activity in response to OT revealed that multiple brain regions are involved in the acceleration information processing. However, as OT create heat, at this stage of the study it is difficult to judge whether is it heat or perceived acceleration that is causing the responses. Further analysis of additional experiments will need to be performed before the nature of vestibular processing will be described with confidence.

Printed by Printforce, the Netherlands